5/11

Write Math!
How to Construct Responses to Open-Ended Math Questions
Level E

Coach™
America's Best for Student Success®

A Haights Cross Communications® Company

Write Math! How to Construct Responses to Open-Ended Math Questions, Level E
100NA
IBSN-10: 1-58620-911-6
IBSN-13: 978-1-58620-911-7

Cover Image: Myron/The Image Bank/Getty Images

Triumph Learning® 136 Madison Avenue, 7th Floor, New York, NY 10016
Kevin McAliley, President and Chief Executive Officer

Printed in the United States of America.

10 9 8

Table of Contents

Letter to the Student

Dear Student,

Welcome to the smart way to write answers to open-ended math questions. You will learn what open-ended math questions are, how to solve them, and how to score them. You will do this by reviewing modeled problems, practicing with guided questions, and answering independent problems. You will work together with your teacher, with your classmates, and with your caregiver at home.

Let's learn to **write math** the smart way!

Have fun!

1. What Is an Open-Ended Math Question?

It is a math problem with a correct answer that you can get to in different ways.

Each way is great as long as:

- **it gets you to the right answer**
- **you show how it got you to the answer**
- **you explain why you chose to answer the question this way**

5-step plan:

1. Read and Think • **2. Select a Strategy** • **3. Solve** • **4. Write/Explain** • **5. Reflect**

This plan will help you answer open-ended questions.

1. Read and Think
2. Select a Strategy
3. Solve
4. Write/Explain
5. Reflect

1. Read and Think

Read the **problem** carefully.

What **question** are you being asked?

- **In your own words, tell what this problem is about.**

What are the **keywords**?

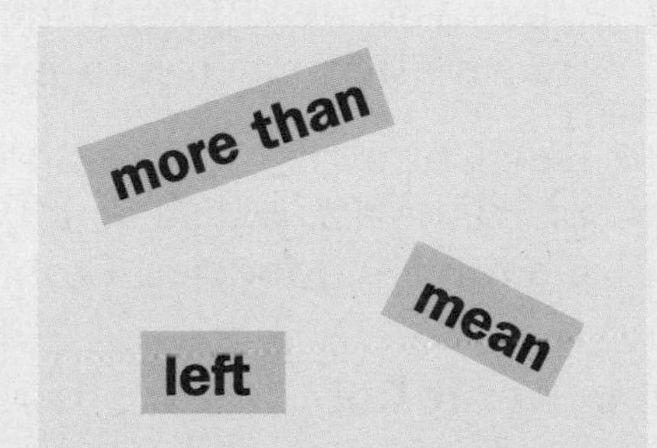

What **facts** are you given?

- **Decide what facts are needed, and which ones are extra.**

2. Select a Strategy

- **How am I going to solve this problem?**
- **What strategy should I use?**

There are lots of strategies to help you solve open-ended math questions. Some are listed below and others you may come up with yourself. In parentheses, you will find places where to locate examples of the strategies in use.

Draw a Picture or Graph . . .

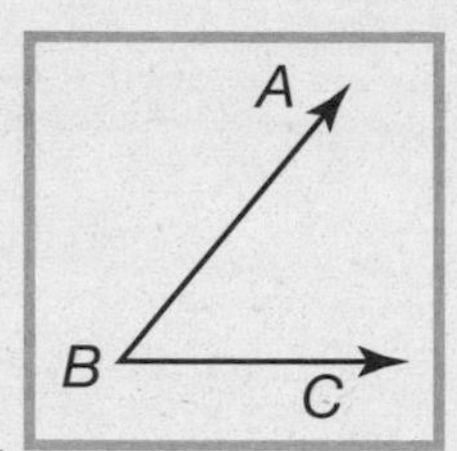

when you need to see the information given in a problem.

(see Chapter 7, Guided Problem #1, *p. 73*, or #2, *p. 77*, or #3, *p. 82*; Chapter 9, Modeled Problem, *p. 107*)

Make a Model or Act It Out . . .

when you need to watch how the solution is found.

(see Chapter 6, Guided Problem #1, *p. 55*)

Make an Organized List or Table . . .

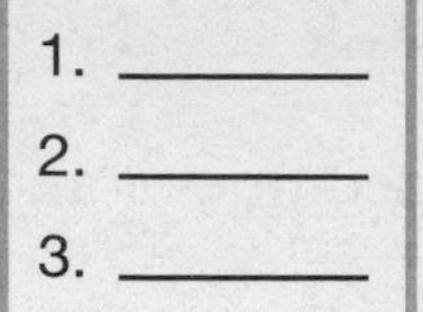

when there is a lot of information scattered throughout a problem. A list or table can help you organize your thinking.

(see Chapter 5, Guided Problem #3, *p. 45*; Chapter 7, Guided Problem #3, *p. 82*; Chapter 8, Guided Problem #1, *p. 93*, or #2, *p. 98*, or #3, *p. 103*; Chapter 9, Guided Problem #2, *p. 116*)

Look For a Pattern . . .

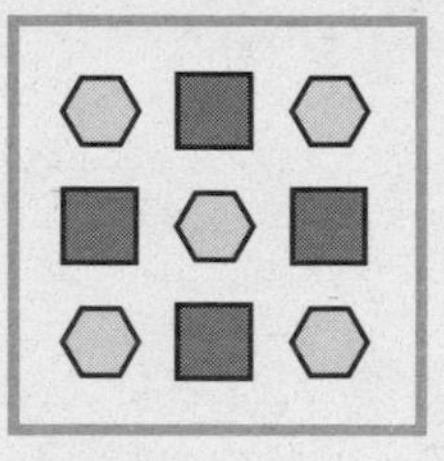

when you need to predict what comes next or find a rule. Making a list or table can often help you find a pattern.

(see Chapter 6, Guided Problem #2, *p. 58–59*)

Guess and Test . . .

when it is difficult to work out the answer to a problem. Make a guess. Then test it. If your guess is incorrect, use that guess to make a better guess.

(see Chapter 5, Guided Problem #1, *p. 36*)

Logical Thinking . . .

when you need to figure out how the information you have fits together.

(see Chapter 6, Guided Problem #2, *pp. 58–59*, or #3, *p. 63*; or Chapter 7, Modeled Problem, *p. 67*; or Chapter 8, Guided Problem #3, *p. 103* or Chapter 9, Modeled Problem, *p. 107*)

Work Backward . . .

when you know the end result or total and need to find a missing part.

(see Chapter 3, Modeled Problem, *p. 19*)

Write a Number Sentence or Algebraic Equation or Use a Formula . . .

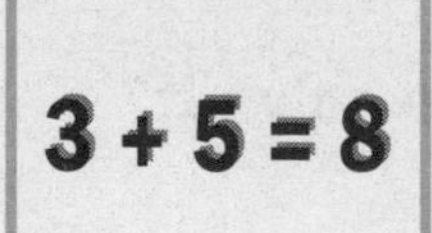

when you need to find a missing amount or show your work.

(see Chapter 6, Modeled Problem, *p. 50*, Guided Problems #1, *p. 55*, or #2, *pp. 58–59*, or #3, *p. 63*; Chapter 9, Guided Problems #1, *p. 112*, or #2, *p. 116*, or #3, *p. 120*)

Divide and Conquer . . .

$3 \times 4 \div 2 =$?

$3 \times 4 = 12$

$12 \div 2 =$ 6

when you must solve more than one equation to find the answer to the main problem. Break down the main question into steps, and solve each step, one at a time. Be sure to tell what each step is. Name the strategy you may use for each step.

(see Chapter 5, Modeled Problem, *pp. 31–32*, Guided Problem #2, *p. 41*; Chapter 8, Modeled Problem, *p. 87*, Guided Problem #1, *p. 93*)

Make It Simpler . . .

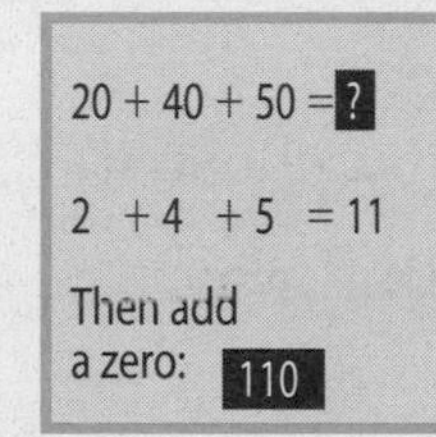

when you must solve a complex problem with large numbers or many items.

Reduce the large numbers to small numbers, or reduce the number of items given.

(see Chapter 5, Guided Problem #2, *p. 41*)

3. Solve

After you pick your strategy, use it to solve the problem.

Use your knowledge of arithmetic and mathematics here.

- **Be very careful—you must get the correct answer. Check your arithmetic!**
- **Label your work. Use units or a sentence that explains the answer.**

4. Write/Explain

Write out an **explanation** of how you solved the problem.

Explain the strategy you chose and why you chose it.

Write your thoughts of why you solved it that way.

Don't leave out any steps.

- **If you came up with a strategy of your own that is not on the list on *pages 8–9*, be sure to explain what the strategy is and why you chose it!**
- **Your writing must be clear. This is very important when you are taking a test. Remember, the person who reads your work must be able to figure out what you did. You can lose points if your writing is not very clear.**

5. Reflect

First, Review Your Work

Use the list below to check your work.

Read It and Think

- ☐ **Did I read the problem at least twice? Do I understand it?**
- ☐ **Did I write down the question being asked?**
- ☐ **Did I write down the keywords in the question?**
- ☐ **Did I write down the facts that are given?**
- ☐ **Did I write down the strategy that I used?**
- ☐ **Did I solve the problem?**
- ☐ **Is my arithmetic correct?**
- ☐ **Did I explain how I solved the problem?**
- ☐ **Did I explain why I chose the strategy and how I used it?**
- ☐ **Did I include all the steps I took to solve it?**
- ☐ **Is my writing clear?**
- ☐ **Did I label my work?**
- ☐ **Does my answer make sense?**
- ☐ **Did I answer the exact question being asked?**

Always check your work!

Then, Improve What You Wrote

How can you improve your writing?

- **Try to rewrite your answer to make it clearer, more accurate, and more complete.**

These 5 steps just reviewed might look like a lot of work. Once you start to use them, they will become very familiar and not seem so hard. You will find that these five steps work with any open-ended math question. This book will help you practice using them.

Working with this book will help you do better on math tests with open-ended questions. You will even learn how to check your answer using the **rubric** on *page 13*.

The Glossary

A Glossary is like a dictionary. It tells the meaning of words. You should learn to use the Glossary found on *pages 145–158*. It contains mathematical words you should know, including **keywords** found in the math problems throughout this book.

Sometimes a word can mean one thing in everyday life, but something else in mathematics. For example, in everyday life, the word **straight** can mean *immediately*, such as "After you wash up, go **straight** to bed." But in math, **straight** can mean *a kind of line that is not bent or curved*, as in, "The sides of a square are four **straight** lines."

If you are not sure what a word means mathematically in this book, look it up in the Glossary.

Tips

- **If you came up with a strategy of your own that is not listed on *pages 8–9*, be sure to explain what the strategy is and why you chose it!**
- **Your writing must be clear. This is very important when you are taking a test.**
- **Remember, the person who reads your work must be able to figure out what you did. You can lose points if your writing is not very clear.**

2. What Is a Rubric?

A **rubric** is a grading system used to score **open-ended math questions**. The person who scores the answers on your test uses a rubric. A rubric can also be used as a guide in answering open-ended questions. It lists things that should be found in your answer.

Throughout this book, you will use the rubric in two ways:

1. to **guide** you in answering an open-ended math question. It will remind you to write a correct, clear, complete, and thoughtful answer.

2. to score yourself as you **double-check** that your answer is complete.

2. What Is a Rubric?

Here is a typical rubric. It is used to score your work from **0** to **4**. **4** is a perfect answer.

4

- You showed you knew what the problem asked.
- You showed you knew what facts were given, including keywords.
- You chose a good strategy and used it correctly.
- Your arithmetic or operations were done correctly.
- You got a correct and complete answer and labeled it.
- You wrote a good, clear explanation of why you chose a strategy and how you used it.
- You put in all the steps you used to get to your answer.
- You explained your thinking clearly.

3

- You showed you knew what the problem asked.
- You showed you knew what facts were given, including keywords.
- You chose a good strategy but may not have used it correctly, OR you may have made an arithmetic error in your work.
- You wrote an explanation of why you chose a strategy and how you used it.
- You might not have used all of the steps to get your answer.
- Your explanation was mostly clear but might not have been entirely complete.

2

- You showed you knew what the problem asked.
- You showed you knew what facts were given, including keywords.
- You chose a good strategy but may not have used it correctly, OR you may have made an arithmetic error in your work.
- Your answer may not be correct.
- Your explanation may not be complete.
- Your explanation may not be clearly written.

1

- You did not understand what the problem asked, OR you did not know what facts were given.
- You did not select a good strategy or did not apply your strategy correctly.
- You made an arithmetic error in your work.
- Your explanation was not complete or you did not write an explanation.
- Your explanation was not clearly written.

0

- You showed no work at all, OR the work you showed had nothing to do with the problem.

Tips

- You will never get a score of **0** if you start the problem.
- You should always write down what you were asked to find out and what facts were given. This shows that you understood the problem, and attempted to solve it.
- Getting used to answering a question using a rubric may seem like a lot of work, but once you start to use it, you will see it as being very helpful. You should practice using a rubric at school and at home.

2. What Is a Rubric?

Let's do a open-ended math question. See how the **5-step plan** fits the rubric to help **YOU** get a **4** on your next answer!

Remember the **5-step plan**:

1. Read and Think • **2. Select a Strategy**

3. Solve • **4. Write/Explain** • **5. Reflect**

1. Read and Think

Read the **problem** carefully.

Modeled Problem

Suzanne and Kevin were taking photographs. Suzanne took 8 rolls of film with 36 pictures in each roll. Kevin took 12 rolls of film with 24 pictures in **each** roll. Who took **more** pictures?

Keywords: each, more

What **question** are we being asked?

- **The question tells you what it is you want to find out. You must answer the question in order to solve the problem correctly.**
- **Here, the question is, "Who took more pictures?"**

What are the **keywords**?

- **each, to consider individually**
- **more, greater in number or amount**

What **facts** are you given?

- **Every problem has facts, data, or information. Facts help you answer the question.**

In this problem, the **facts** are:

- **Suzanne took 8 rolls of film with 36 pictures in each roll.**
- **Kevin took 12 rolls of film with 24 pictures in each roll.**

2. Select a Strategy

In order to solve a problem, you need to use a **strategy**.

There are many **strategies** you can use. *Chapter 1, pages 8–9* shows some strategies you might use. You may also choose to use one of your own.

Let's look at how two students chose different strategies to solve this problem.

First Solution

First, we will show what Todd knows about arithmetic.

This involves using a strategy, **Writing a Number Sentence.**

3. Solve

First solution—**Write a Number Sentence**:

Suzanne's photographs
8 rolls with 36 pictures in each roll
8 × 36 = 288

Kevin's photographs
12 rolls with 24 pictures in each roll
12 × 24 = 288
Suzanne and Kevin took the same number of photographs.

Turn the page to see the end of this problem.

Second Solution

Second, we'll review what Carolyn did using a strategy called **Divide and Conquer**.

3. Solve

Second solution—**Divide and Conquer**:

First, I found how many photographs Suzanne took.

```
  36 pictures in each roll
× 8 rolls of film
 248
```

Then, I found how many photographs Kevin took.

```
   24 pictures in each roll
×  12 rolls of film
   48
+ 240
  288
```

Finally, I compared the numbers.
248 < 288, so Kevin took more photographs.

Tip

- Many problems can be solved using different strategies. As long as your choice leads to a correct answer and a correct explanation, it is a good choice!

First Solution

4. Write/Explain

You must give a written explanation of how you solved the problem and what you were thinking.

Clearly explain what you did and why you did it.

Do not leave out any steps.

> I multiplied 8 × 36 to find the number of photographs Suzanne took. I found that she took 288 photographs. Then I multiplied 12 × 24 to find the number of photographs Kevin took. He also took 288 photographs. They each took 288 photographs. They took the same number of photographs.

5. Reflect

Todd reviewed his work by checking it against the rubric. He answered the problem being asked, chose a good strategy and used it correctly. His arithmetic was correct. He wrote a complete and clear answer to explain why he chose the strategy and how he solved the problem. Finally, he labeled his work.

Score

Todd gets a score of **4**.

Second Solution

4. Write/Explain

> I multiplied 36 × 8 to find how many photographs Suzanne took. Then I multiplied 24 × 12 to find how many photographs Kevin took. Suzanne took 248 photographs and Kevin took 288 photographs, so Kevin took more photographs.

5. Reflect

Carolyn reviewed her work by checking it against the rubric. She answered the problem being asked. Carolyn chose a good strategy and used it correctly. She wrote a complete and clear answer and she labeled her work. However, Carolyn did not catch her arithmetic error. Her product for the number of photographs that Suzanne took is incorrect.

Score

Carla gets a score of **3**.

She could improve her work to get a **4** by multiplying 36 × 8 = 288 and remembering to regroup when multiplying.

Using the Rubric

Whenever you solve an open-ended math question, you should use the rubric on *page 13* as a guide. Make a copy of it from the book. Place the rubric next to the question you are working on. Use the list in the score of **4** box as a checklist. This will remind you what to include in your answer for the highest score possible.

Reviewing Your and a Partner's Work

After you finish solving the question, **self-assess**. This means that you should use the whole rubric to review your work and score. How well did you do? If you need to raise your score, take the time to do so.

You may also **peer-assess**. Swap your work with a partner. Use the rubric to score each other's solution. Now talk about the different answers and the scores that were given. What might seem clear to you may not be clear to your friend. Partners can help each other learn what should be improved. You can discuss different ways to solve the same problem. The more you talk about your mathematics, the more you will understand how to improve your work.

2-Step Decision-Making Process

Some students find it easier before scoring with a rubric, to first use the **2-Step Decision-Making Process** as seen in the next column. It helps decide if your or your partner's answer is a **3** or **4** or a **1** or **2**.

2-Step Decision-Making Process

Before you use a rubric, use a **2-Step Decision-Making Process**. This will give you a jump on the scoring of your work or your partner's work.

Decide if your work is:

- ***acceptable* (3 or 4)**

  or

- ***unacceptable* (1 or 2).**

If your work is ***acceptable***, decide if it is:

- **full and complete (4)**

  or

- **nearly full and complete, but not perfect (3).**

If your work is ***unacceptable***, decide if it shows:

- **limited or only some understanding (2)**

  or

- **little or no understanding (1).**

A **(0)** is **no attempt**.

Solve questions using the rubric as a guide. You will see improvement in your answers and ability. In time you may no longer need a rubric to guide you. The information that must be included in your answer will become very familiar. Then, you will only need the rubric to score your answer.

3. How to Answer an Open-Ended Math Question

We know what an **open-ended math problem** is. We know how to solve it and how it will be scored. Now let's take a problem and solve it together. Then we will see how other students answered it. We will use the **rubric** to see how well they did. Then we will talk about how they could **improve** their answers.

Modeled Problem

Joe is 3 **years older** than Darlene. Darlene is three **times** as **old** as Tom. Tom is 4 **years younger** than Carlos. Carlos is 8 **years** old. How **old** is Joe?

Keywords: years, older, times, old, younger

1. Read and Think

What **question** are we asked?

- **How old is Joe?**

What are the **keywords**?

- **years, older, times, old, younger**

What **facts** are we given?

- **Joe is 3 years older than Darlene.**
- **Darlene is 3 times as old as Tom.**
- **Tom is 4 years younger than Carlos.**
- **Carlos is 8 years old.**

What is **going on**?

- **We know that Carlos is 8 years old.**
  Other than that, we don't know anyone's ages.

2. Select a Strategy

Let's **Work Backward** to find Joe's age.
Working backward is one of our strategies.

First, let's use Carlos's age to find Tom's age.
Then let's use Tom's age to find Darlene's age.
Finally, we can use Darlene's age to find Joe's age.

3. Solve

Use Carlos's age to find Tom's age: $8 - 4 = 4$.
Tom is 4 years old.

Use Tom's age to find Darlene's age: $4 \times 3 = 12$.
Darlene is 12 years old.

Use Darlene's age to find Joe's age: $12 + 3 = 15$.
Joe is 15 years old.

4. Write/Explain

We used the **Work Backward** strategy to find Joe's age. We started by using Carlos's age to find Tom's age. Then we used Tom's age to find Darlene's age and Darlene's age to find Joe's age. Joe is 15 years old.

5. Reflect

We reviewed our work. We picked a good strategy and used it correctly. The math is correct. It corresponds to the information given in the question.

Score

This solution would earn a perfect **4** on our rubric.

- **We showed that we knew what was asked and what information was given, including facts and keywords.**
- **We chose a good strategy and applied it correctly.**
- **We did the arithmetic and got the correct answer.**
- **We wrote a good explanation of how and why we used our strategy.**
- **We included all of our steps.**
- **We labeled our work.**
- **We clearly explained what we did and why.**

Now let's look at some solutions that were done by other students.

Troy's Paper

Question: How old is Joe?
Keywords: years, times
Facts: Joe is 3 years older than Darlene.
Darlene is 3 times as old as Tom.
Tom is 4 years younger than Carlos.
Carlos is 8 years old.
Strategy: Work Backward
Solve: Carlos is 8 years old.
Tom is $8 + 4 = 12$ years old.
Darlene is $12 \times 3 = 36$ years old.
Joe is $36 + 3 = 39$ years old.

Let's use our rubric to see how well Troy did.

- Did he show that he knew what the problem asked? **Yes.**
- Did he know what the keywords were? **No. He missed "older" and "younger."**
- Did he show that he knew what facts were given? **Yes.**

- Did he name and use the correct strategy? **Yes. He Worked Backward.**
- Was his math correct? **No. If Tom is 4 years younger than Carlos, the proper equation to set up to find Tom's age is 8 – 4 = 4.**
- Did he label his work? **Yes.**
- Was his answer correct? **No. He made a mistake in Tom's age, which led to incorrect answers.**
- Were all of his steps included? **Yes. He found everyone's ages.**
- Did he explain why he chose the strategy and how it was used? **No.**
- Did he write a good, clear explanation of his work? **No.**

Score

Troy would get a **2** on our rubric.

He knew what was asked. He knew the facts, but he didn't identify all of the keywords, which led to incorrect mathematics and answers. He picked a correct strategy and use it correctly. He labeled his work and included all his steps. He did not write a complete, clear explanation of his work. He did not explain how he used the strategy.

To get a 4, Troy needs to check his work and pay attention to all of the keywords. If he had, it is possible that he would have seen his error and corrected it to find the correct ages for Tom, Darlene, and Joe.

Wu's Paper

Question: How old is Joe?
Keywords: years, older, times, old, younger
Facts: Joe is 3 years older than Darlene.
Darlene is 3 times as old as Tom.
Tom is 4 years younger than Carlos.
Carlos is 8 years old.
Strategies: I Wrote a Number Sentence, and I Worked Backward.
Solve:

Joe's age = (8 − 4) × 3 + 3
Joe's age = 4 × 3 + 3
Joe's age = 12 + 3
Joe's age = 15
Joe is 15.

Write/Explain: I Wrote a Number Sentence and Worked Backward. I started with Carlos's age and used the information given about everyone's age to find Joe's age. I determined that Joe is 15 years old.

- Did he show that he knew what the problem asked? **Yes.**
- Did he know what the keywords were? **Yes.**
- Did he show that he knew what facts were given? **Yes.**
- Did he name and use the correct strategy? **Yes.**
- Was his math correct? **Yes.**
- Did he label his work? **Yes.**
- Was his answer correct? **Yes.**
- Were all of his steps included? **Yes.**
- Did he explain why he chose the strategy and how it was used? **Yes.**
- Did he write a good, clear explanation of his work? **Yes.**

Score

Wu would receive a **4** on our rubric.

Courtney's Paper

Question: How old is Joe?

Keywords: years, older, times, old, younger

Facts: Joe is 3 years older than Darlene.
Darlene is 3 years older than Tom.
Tom is 4 years younger than Carlos.
Carlos is 8 years old.

Strategy: Work Backward

Solve:
Carlos is 8.
Tom is 4 years younger than Carlos: $8 - 4 = 4$.
Darlene is 3 years older than Tom: $4 + 3 = 7$.
Joe is 3 years older than Darlene: $7 + 3 = 10$.
Joe is 10 years old.

Write/Explain: I Worked Backward to find Joe's age.

- Did she show what the problem asked? **Yes.**
- Did she know what the keywords were? **Yes.**
- Did she show that she knew what facts were given? **No. Darlene is 3 times as old as Tom, not 3 years older.**
- Did she name and use the correct strategy? **Yes.**

- Was her math correct? **Yes.**
- Did she label her work? **Yes.**
- Was her answer correct? **No. Darlene's age and Joe's age are incorrect.**
- Were all of her steps included? **Yes.**
- Did she explain why she chose the strategy and how it was used? **No. She said what strategy she used, but did not explain how she used it.**
- Did she write a good, clear explanation of her work? **No.**

Score

Courtney would receive a **2** on our rubric.

To get a 4, Courtney would need to know what *times* means. If she had multiplied by 3 instead of adding 3, she would have correctly found Darlene's age and then Joe's age. Courtney would also need to explain how she used the **Work Backward** strategy. She also needs to give a clear explanation of her work.

Travis's Paper

Tom is 4 years old.
Darlene is 12 years old.
Joe is 15 years old.
Write/Explain: I used Tom's age to find Darlene's age to find Joe's age. Joe is 15 years old.

Travis did find the correct answer, but it was probably a lucky guess. He did not use the strategy of **Working Backward** correctly, as he did not use the information about Carlos. He also did not provide the keywords, the facts, or the question that was asked. His explanation was not complete.

Score

Travis would receive a **1** on our rubric.

To get a 4, Travis needs to give the question that was asked, the keywords, and the facts, use the strategy he chose correctly, and then write a detailed explanation of his work.

4. How NOT to Get a ZERO!

No one wants to get a zero on an open-ended math problem. However, you can almost always get some points. The only person who gets a **0** is the person who **leaves the paper blank** or who **writes something that doesn't have anything to do with the problem**. Let's see how we can start by scoring a **1 or 2** on our work, and then bring it up to a **3 or 4**.

And remember:

1. **Read and Think**
2. **Select a Strategy**
3. **Solve**
4. **Write/Explain**
5. **Reflect**

4. How NOT to Get a ZERO!

How to Get a 1 or 2!

Here is how to get **some credit** on an open-ended math question.

1. **Read** the question. Then **reread** it.

   Ask: "What are the **keywords** to help you solve the problem?

   Finish the sentence: **"The keyword/s are**

   ________________________________."

   You will get credit for listing the keywords.

2. **Understand** the problem. **Repeat the story** of the problem in your own words.

   Ask: "What am I **being asked** to do? What do I **need to find**?"

   Finish the sentence: **"I need to find**

   ________________________________."

   You will get credit for listing what was asked.

3. **Find the facts** in the problem.

   Ask: "What does the problem tell me? What do I **know**?"

   Finish the sentence: **"The things I know are**

   ________________________________."

   You will get credit for writing the facts.

4. Figure out what **strategy** you will use to help you solve the problem.

   Ask: "What **can help** me to find what I need to know?"

   Finish the sentence: **"The strategy I will use is**

   ________________________________."

   You will get credit for listing the strategy you use.

Tip

- To get points right away, always begin by writing down what you are asked to find and what you are given as facts.

How to Get a 1 or 2!

You will practice these steps as you help solve the modeled problem introduced on the following page.

(continued on the next page.)

4. How NOT to Get a ZERO!

Let's do an open-ended math problem together. First, let's try to get a **1 or 2.**

Modeled Problem

Carrie bought a DVD for $19.95 and a CD for $13.99. She also bought a sweater for $35.29. How **much more** did she spend on the sweater than on the DVD and CD **combined**?

Keywords: much, more, combined

1. Read and Think

1. Carefully **read this question. Reread** the question to fully understand it.
2. What **question** are you being asked?

- **I need to find how much more the sweater cost than the DVD and CD combined.**

By writing what question you are being asked, you can get a score of a 1 or 2.

3. What are the **keywords** to help you solve the problem?

- **much, more, combined**

By listing the keywords, you can get a score of a 1.

4. What are the **facts**?

- **Carrie bought a DVD for $19.95.**
- **Carrie bought a CD for $13.99.**
- **Carrie bought a sweater for $35.29.**

By listing the facts, you can get a score of a 1 or 2.

Hint

To get points right away, always begin by writing down **what you are asked** and what are given as **facts**.

2. Select a Strategy

Now you have to pick a strategy and solve the problem. There are lots of strategies from which to pick. You may choose one from *pages 8–9* or choose one of your own. Your classmate may choose a different strategy than you to solve the same problem. You may solve it differently, but you both can get the right answer. There is not only one right way.

1. What **strategy** will you use?

- **I will use a strategy called Divide and Conquer.**

***First,* I will Make a Table to organize all the facts in the problem. *Then,* I will Write Number Sentences to help me solve the problem and find the answer.**

By writing what strategy you chose you can get a score of a 1 or 2.

How to Get a 3 or 4!

Now, let's try to increase our score on the same problem from a **1 or 2** to a **3 or 4.**

First, let's review. ***Remember, you can always get some credit*** for listing the keywords and the question that is asked. You will also receive points by writing the facts that are given. Finally, credit will be given for listing the strategy you have chosen. By doing this, you will receive a score of at least a **1 or 2. Now it is time to raise your score to a 3 or 4.** Use all the following information introduced to do so.

So here we go:

Here is how to take a score of **1 or 2** and make it a **3 or 4.** We will continue using the same modeled problem.

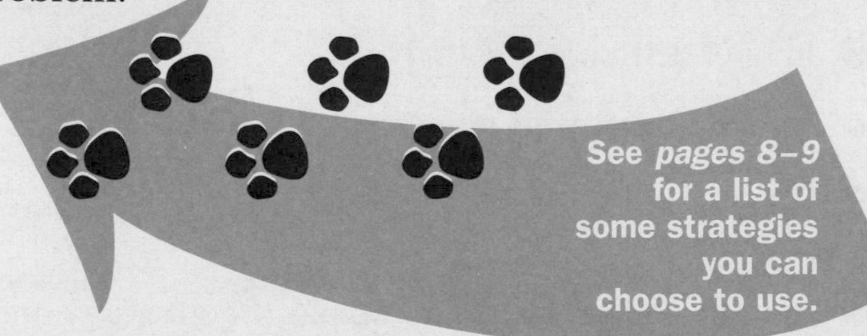

See *pages 8–9* for a list of some strategies you can choose to use.

3. Solve

Go back to the original problem. How can we find how much more Carrie spent on the sweater than on the DVD and CD combined?

First, we **Write a Number Sentence** to find the cost of the DVD and CD combined.

$19.95
+ 13.99
―――――
$33.94

The cost of the DVD and the CD is $33.94. *Now,* we **Write another Number Sentence** to find how much more the sweater cost than the DVD and CD combined.

$35.29	Cost of sweater
– 33.94	Cost of DVD and CD combined
$1.35	How much more Carrie spent on the sweater

Tip

- We're going to use a lot of strategies in this book.
- An open-ended math problem can be solved in more than one way. And if your way works out and gives the correct answer, you are right!

4. How NOT to Get a ZERO!

Always remember to **label** your work. Use **units** or a sentence that explains what you found.

4. Write/Explain

The person marking your paper does not know what you were thinking. You must explain why you chose the strategy. You should explain how you solved the problem. Your work should be labeled. Don't leave out any steps. Be sure to reread your writing, making sure your work is clear and complete.

I used the **Divide and Conquer** strategy. The question asks how much more Carrie spent on the sweater than on the CD and DVD combined, so first I need to find the combined price of the CD and DVD. Then I need to subtract this amount from the cost of the sweater to find the difference.

First find the cost of the DVD and CD.

$$\begin{array}{r} \$19.95 \\ +\ 13.99 \\ \hline \$33.94 \end{array}$$

Then subtract $33.94 from the cost of the sweater, $35.29, to find how much more the sweater cost than the DVD and CD combined.

$$\begin{array}{r} \$35.29 \\ -\ 33.94 \\ \hline \$1.35 \end{array}$$

$1.35 is how much more Carrie spent on the sweater than on the CD and DVD combined.

5. Reflect

Review Your Work and Improve It!

After you solve the problem, carefully review your work.

- **Did you write what the problem asked you to find?**
- **Did you list all the facts, including keywords?**
- **Did you list the strategy you chose to use?**

If you did these things, you will get a score of 1 or 2.

- **Did you use the right strategy?**
- **Is your arithmetic right?**
- **Did you label your work?**
- **Did you write out all the steps to solving the problem?**
- **Did you explain why you chose the strategy and how you used it?**

- Did you explain why you solved the problem the way you did?
- Is your writing clear?

If you did these things, you will raise your score from a 1 or 2 to a 3 or 4.

If you do all the things we have suggested, you **CANNOT** get a **0.**

Remember, never leave your paper blank.

Working with Peers

You might want to exchange papers with a friend in class. See if your friend understands what you wrote. That's a good way to see how clearly you explained your work.

Here is a checklist for you to follow. It will make sure you have done your best job. Keep practicing what is on this list. You will improve solving open-ended mathematics questions.

Use a Checklist and Your Rubric

Here is a checklist that will help you make sure you have done the best that you can. You can use this checklist and the rubric that was in Chapter Two to improve your work.

Read It and Think

- ☐ Did I read the problem at least twice? Do I understand it?
- ☐ Did I write down the question being asked?
- ☐ Did I write down the keywords in the problem?
- ☐ Did I write down the facts that are given?
- ☐ Did I write down the strategy that I used?
- ☐ Did I solve the problem?
- ☐ Is my arithmetic correct?
- ☐ Did I explain how I solved the problem?
- ☐ Did I explain why I chose the strategy and how I used it?
- ☐ Did I include all the steps I took to solve it?
- ☐ Is my writing clear?
- ☐ Did I label my work?
- ☐ Does my answer make sense?
- ☐ Did I answer the exact question being asked?

Always check your work!

5. Number and Operations

Numbers are all around us, every day. They make up our **phone number,** tell us the **location** of where we live, let us know **how much** something costs, and on and on! We are going to learn more about numbers, explaining how you use them. We will understand using one number alone, how numbers can be used together, how they can be compared, and more.

Here is a problem that you might have to solve on a test. Let's solve it together to show what a model answer, a score of **4,** might look like. Then we will check our solution using a **rubric**.

Modeled Problem

Sarah is a high jumper. She raises the bar 3 inches **after each** successful jump. If the bar starts at 3 feet 8 inches, and she has 3 successful jumps, how **high** is the bar after Sarah has her third successful jump?

Keywords: after, each, high

1. Read and Think

What **question** are we being asked?

- **We are asked to find out what height the bar was after Sarah's third successful jump.**

What are the **keywords**?

- **after, each, high**

What are the **facts**?

- **The bar starts at 3 feet 8 inches.**
- **The height of the bar increases 3 inches after each jump.**
- **Sarah has 3 successful jumps.**

2. Select a Strategy

There are *two* parts to this problem. The *first* part is *multiplication*. The *second* part is *addition*. We will use the **Divide and Conquer** strategy.

3. Solve

The first part is to find how many inches the height of the bar increased.

3 jumps × 3 inches = 9 inches

The height of the bar increased by 9 inches.

The second part is to find the height of the bar.

3 feet 8 inches
+ 9 inches
3 feet 17 inches = 4 feet 5 inches

(Remember, there are 12 inches in 1 foot, so 12 of those 17 inches are the fourth foot. 17 inches minus 12 inches leaves 5 inches extra.)

4. Write/Explain

To solve this problem, we had to find how high the bar was after Sarah completed her third successful jump. It took two steps to do this, so we used the **Divide and Conquer** strategy. For the first part of this strategy, we multiplied 3 jumps by 3 inches to figure out how much the height increased. It increased by 9 inches. Then we had to add this amount to the height that the bar started at to find how high it was after 3 successful jumps. It was 3 feet 17 inches high, which is 4 feet 5 inches.

5. Reflect

Let's review our work and answer.

- Did we show that we knew what the problem asked? **Yes. We answered the question that was asked.**
- Did we know what the keywords were? **Yes.**
- Did we show that we knew what facts were given? **Yes.**
- Did we name and use the correct strategy? **Yes.**
- Was our math correct? **Yes. We checked it. It was correct.**
- Did we label our work? **Yes.**
- Was our answer correct? **Yes.**
- Were all of our steps included? **Yes.**
- Did we explain why we chose the strategy and how it was used? **Yes.**
- Did we write a good, clear explanation of our work? **Yes.**

Score

This solution would earn a **4** on our rubric. It is perfect.

On the following pages are some **Guided Open-Ended Math Problems.**

For each problem there are **four parts**. In the **first part**, you will solve the problem with guided help. In the **second part**, you will score and correct a solution with guided help. The **third part** shows one solution that scores a perfect **4**. This solution may or may not differ from your way. The **fourth part** has answers to the **first** and **second parts** so you can check your work.

Guided Problem #1

While on vacation, Juan spent $4.56 on stamps for letters and postcards to his friends. Stamps for letters cost $0.39 each and stamps for postcards cost $0.24 each. Juan wrote to 14 friends. How many letters did he send? How many postcards did he send?

Keywords: ? ?

1. Try It Yourself.

Answer the questions below to get a score of **4**.

What **question** are you being asked?

____

____

What are the **keywords**?

____

____

What are the **facts** you need to solve the problem?

____

____

What **strategy** can you use to solve the problem?

____

____

____

> **Hint**
>
> Possible answers include **Guess and Test** and **Look For a Pattern.**

Solve the problem.

____

____

____

____

____

Write/Explain what you did to solve the problem.

____

____

____

____

____

Reflect. Review and improve your work.

____

____

____

____

2. Dennis Tries It.

Dennis's Paper

Questions: How many letters did Juan send?
How many postcards did Juan send?
Keywords: each, many
Facts: Stamps for letters cost $0.39 each.
Stamps for postcards cost $0.24 each.
Juan wrote to 14 friends.
He spent $4.56.
Strategy: I can use Guess and Test.
Solve:

Letters	Cost	Postcards	Cost	Total Cost
7	$2.73	7	$1.68	$4.41
8	$3.12	6	$1.44	$4.56

Juan sent 8 postcards and 6 letters.
Write/Explain: I used Guess and Test to find how many postcards and how many letters Juan sent.
I multiplied the number of postcards times $0.24 and the number of letters times $0.39.
The sum of my products is equal to $4.56.

Score the Answer.

According to the rubric, from **1** to **3** what score would you give Dennis? Explain why you gave that score.

____

____

____

____

____

Make it a 4! Rewrite.

____

____

____

____

____

Use the rubric on *page 13* to score this work.

3. Carl Tries It.

Remember, there is often more than one way to solve a problem. Here is how Carl solved this problem.

Carl's Paper

Questions: How many letters did Juan send?
How many postcards did Juan send?
Keywords: each, many
Facts: Stamps for letters cost $0.39 each.
Stamps for postcards cost $0.24 each.
Juan wrote to 14 friends.
He spent $4.56.
Strategies: Logical Thinking, Guess and Test
Solve:
I divided 456 ÷ 14 = 32 R8.
Since 32 is closer to 39 than 24, Juan sent more letters than postcards.
I tried 8 letters and 6 postcards.
$0.39 × 8 = $3.12
$0.24 × 6 = $1.44
$3.12 + $1.44 = $4.56.
Juan sent 8 letters and 6 postcards.

Write/Explain: I used Logical Thinking to find the mean of the cost of the stamps by dividing 456 by 14, which was 32 R8. Since 32 is closer to 39 than 24, Juan sent more letters than stamps. I then used Guess and Test to find that Juan sent 8 letters and 6 postcards.

Score: Carl's solution would earn a **4** on a test. He identified the question that was asked, the keywords, and the facts. He correctly picked and used a strategy correctly. Then he clearly explained the steps taken to solve the problem. He also labeled his work.

4. Answers to Parts 1 and 2.

Guided Problem #1

While on vacation, Juan spent $4.22 on stamps for letters and postcards to his friends. Stamps for letters cost $0.37 **each** and stamps for postcards cost $0.21 **each.** Juan wrote to 14 friends. How **many** letters did he send? How **many** postcards did he send?

Keywords: ? ?

1. Try It Yourself. (page 33)

Questions: How many letters did Juan send? How many postcards did Juan send?

Keywords: each, many

Facts: Stamps for letters cost $0.37 each.
Stamps for postcards cost $0.21 each.
Juan wrote to 14 friends.
He spent $4.22.

Strategy: Guess and Test

Solve:

Postcards	Cost	Letters	Cost	Total Cost
7	$2.59	7	$1.47	$4.06
8	$2.96	6	$1.26	$4.22

Write/Explain: I used the **Guess and Test** strategy. I started with an equal number of letters and postcards. Since the cost was less than $4.22, I added another letter and subtracted a postcard. That guess led me to the correct answer of 8 letters and 6 postcards.

2. Dennis Tries It. (page 34)

Score the Answer: I would give Dennis a **3**. Dennis wrote down the question asked, gave the keywords, and listed the facts. He named and used a correct strategy. His math was right and was labeled. However, Dennis mixed up the letters and the postcards in his table and therefore his answer was incorrect. All of his steps were included. He explained how he used his strategy and his explanation was clear.

Make it a 4! Rewrite.

Postcards	Cost	Letters	Cost	Total Cost
7	$2.59	7	$1.47	$4.06
8	$2.96	6	$1.26	$4.22

I multiplied the number of letters times $0.37 and the number of postcards times $0.21. The sum of my products is equal to $4.22. Juan sent 8 letters and 6 postcards.

Guided Problem #2

Gareth and Jennifer plan to hike $9\frac{1}{4}$ miles today. They have hiked $5\frac{1}{2}$ miles so far. How many more miles have they already hiked than they still have to hike?

Keywords: ? ?

1. Try It Yourself.

Answer the questions below to get a score of **4.**

What **question** are you being asked?

____

____

____

____

What are the **keywords**?

____

____

____

____

What are the **facts** you need to solve the problem?

____

____

____

____

What **strategy** can you use to solve the problem?

____

____

____

Hint

Possible answers include **Divide and Conquer** and **Write an Equation.**

Solve the problem.

____

____

____

____

____

____

____

Write/Explain what you did to solve the problem.

____

____

____

____

____

____

____

____

Reflect. Review and improve your work.

____

____

____

____

____

____

____

____

2. Len Tries It.

Len's Paper

Question: How many more miles have they hiked than they still have to hike?

Keywords: miles, many more, far

Facts: They have hiked $5\frac{1}{2}$ miles.

They plan on hiking $9\frac{1}{4}$ miles.

Strategy: Divide and Conquer

Solve:

$$\begin{array}{rcl} 9\frac{1}{4} & = & 9\frac{5}{4} \\ -\ 5\frac{1}{2} & = & 5\frac{2}{4} \\ \hline & & 4\frac{3}{4} \end{array}$$

$$\begin{array}{rcl} 5\frac{2}{4} & = & 4\frac{6}{4} \\ -\ 4\frac{3}{4} & = & 4\frac{3}{4} \\ \hline & & \frac{3}{4} \end{array}$$

Gareth and Jennifer have hiked $\frac{3}{4}$ mile more than what they still have to hike.

Write/Explain: I used the Divide and Conquer strategy. I found out how many miles they still have to hike. Then I found how many more miles they have already hiked more than they still have to hike.

Score the Answer.

According to the rubric, from **1** to **3** what score would you give Lew? Explain why you gave that score.

Make it a 4! Rewrite.

3. Josh Tries It.

Remember, there is often more than one way to solve a problem. Here is how Josh solved this problem.

Josh's Paper

Question: How many more miles have they already hiked than they still have to hike?

Keywords: miles, many more, far

Facts: Gareth and Jennifer plan to hike $9\frac{1}{4}$ miles. They have hiked $5\frac{1}{2}$ miles.

Strategy: I Wrote Equations.

Solve:

$9\frac{1}{4} - 5\frac{1}{2} = m$

$5\frac{1}{2} - m = x$

m = the number of miles that Gareth and Jennifer still have to hike.

x = the number of miles more that Gareth and Jennifer have hiked than they still have to hike.

$$9\frac{1}{4} = 8\frac{5}{4}$$

$$-\,5\frac{1}{2} = 5\frac{2}{4}$$

$$3\frac{3}{4} = m$$

$$5\frac{2}{4} = 4\frac{6}{4}$$

$$-\,3\frac{3}{4} = 3\frac{3}{4}$$

$$1\frac{3}{4} = x$$

Gareth and Jennifer have hiked $1\frac{3}{4}$ more miles than they still have to hike.

Write/Explain: I Wrote an Equation to find out how many miles Gareth and Jennifer still had to hike. Then I wrote another equation to find how many more miles they have hiked than they still have to hike. I used the difference from the first equation as the number that was subtracted from the miles they have already hiked.

Score: Josh's solution would earn a **4** on our rubric. He identified the question that was asked, the keywords, and the facts, and correctly picked and used a strategy. He clearly explained the steps taken to solve the problem. Josh labeled his work.

4. Answers to Parts 1 and 2.

Guided Problem #2

Gareth and Jennifer plan to hike $9\frac{1}{4}$ miles today. They have hiked $5\frac{1}{2}$ miles so far. How many more miles have they already hiked than they still have to hike?

Keywords: ? ?

1. Try It Yourself. (pages 37–38)

Question: How many more miles have they hiked than they still have to hike?

Keywords: miles, many more, far

Facts: Gareth and Jennifer plan to hike $9\frac{1}{4}$ miles. They have hiked $5\frac{1}{2}$ miles.

Strategies: Divide and Conquer or Make It Simple

Solve: Miles they still have to hike:

$$\begin{array}{rcl} 9\frac{1}{4} & = & 8\frac{5}{4} \\ -\ 5\frac{1}{2} & = & 5\frac{2}{4} \\ \hline & & 3\frac{3}{4} \end{array}$$

Miles more they have hiked than they still have to hikc:

$$\begin{array}{rcl} 5\frac{2}{4} & = & 4\frac{6}{4} \\ -\ 4\frac{3}{4} & = & 3\frac{3}{4} \\ \hline & & 1\frac{3}{4} \end{array}$$

Gareth and Jennifer have hiked $1\frac{3}{4}$ miles more than they still have to hike.

To solve a simpler problem, change functions to decimals and subtract.

$9\frac{1}{4} = 9.25 \quad 5\frac{1}{2} = 5.50$

$9.25 - 5.50 = 3.75$

$5.50 - 3.75 = 1.75$

Change 1.75 to $1\frac{3}{4}$.

Write/Explain: I used the **Divide and Conquer** strategy. *First,* I **Wrote a Number Sentence.** I subtracted the number of miles they have hiked from the number of miles they plan to hike to find how many more miles they still have to hike. *Then,* I **Wrote another Number Sentence**. I subtracted the number of miles they still have to hike from the number of miles they have hiked. They have hiked $1\frac{3}{4}$ more miles than they still have to hike.

I could also have **Made It Simpler**. I would change the fractions to decimals so it is easier to subtract. The answer is 1.75, so I would change that back to a fraction of $1\frac{3}{4}$ miles.

2. Len Tries It. (pages 38–39)

Score the Answer: Len would get a **3** because his mathematics is wrong. He forgot to subtract 1 from the whole-number part when he regrouped $9\frac{1}{4}$. This error caused his answer to be incorrect. He also forgot to label his answer as miles. However, he knew what the question asked, he gave the keywords and facts, and he picked a good strategy and used it correctly. All of his steps were included and his explanation was clear.

Make it a 4! Rewrite.

Miles they still have to hike:

$$\begin{array}{rcl} 9\frac{1}{4} & = & 8\frac{5}{4} \\ -\ 5\frac{1}{2} & = & 5\frac{2}{4} \\ \hline & & 3\frac{3}{4} \end{array}$$

Miles more they have hiked than still have to hike:

$$\begin{array}{rcl} 5\frac{2}{4} & = & 4\frac{6}{4} \\ -\ 3\frac{3}{4} & = & 3\frac{3}{4} \\ \hline & & 1\frac{3}{4} \end{array}$$

Gareth and Jennifer have already hiked $1\frac{3}{4}$ miles more than they still have to hike.

Guided Problem #3

Lizzie left her home at 7:45 a.m. to go to the beach with friends. She returned home at 6:30 p.m. For how long was Lizzie away with her friends?

Keywords: ? ?

1. Try It Yourself.

Answer the questions below to get a score of **4.**

What **question** are you being asked?

____

____

____

What are the **keywords?**

____

____

____

What are the **facts** you need to solve the problem?

____

____

____

What **strategy** can you use to solve the problem?

____

____

Hint

Possible answers include **Make a Table** and **Divide and Conquer**.

Solve the problem.

____

____

____

Write/Explain what you did to solve the problem.

____

____

____

Reflect. Review and improve your work.

____

____

____

Use the rubric on *page 13* to score this work.

2. Buddy Tries It.

Buddy's Paper

Question: For how long was Lizzie away?
Keywords: left, returned, long
Facts: She left at 7:45 a.m. She returned at 6:30 p.m.
Strategy: Divide and Conquer
Solve:
7:45 to 7:30 is 15 minutes.
7:30 to 6:30 is 1 hour.
1 hour + 15 minutes = 1 hour 15 minutes
Lizzie was away for 1 hour 15 minutes.
Write/Explain: I used the Divide and Conquer strategy to find how long Lizzie was away. First, using number sentences, I found the number of minutes she was away. Then I used another number sentence to find the number of hours Lizzie was away. She was away 1 hour 15 minutes.

Score the Answer.

According to the rubric, from **1** to **3** what score would you give Buddy? Explain why you gave that score.

____

____

____

____

Make it a 4! Rewrite.

____

____

____

____

____

____

3. Skylar Tries It.

Remember there is often more than one way to solve a problem. Here is how Skylar solved this problem.

Skylar's Paper

Question: How long was Lizzie away?

Keywords: left, returned, long

Facts: Lizzie left at 7:45 a.m.

Lizzie returned at 6:30 p.m.

Strategy: Make It Simpler

Solve:

```
  12:00
-  7:45
   4:15
```

4:15 + 6:30 = 10:45

Lizzie was away for 10 hours 45 minutes.

Write/Explain: I Made It Simpler. I subtracted the time when Lizzie left from 12 noon to find that she was away for 4 hours 15 minutes at that point. Then I added 6 hours 30 minutes (the amount of time she was away from noon) to the difference to find she was away for a total of 10 hours 45 minutes.

Score: Skylar earned a perfect **4** on our rubric. She identified the question that was asked, the keywords, and the facts, and picked a good strategy. She explained how she used the strategy and clearly explained the steps she used to solve the problem. She also labeled her work.

4. Answers to Parts 1 and 2.

Guided Problem #3

Lizzie left her home at 7:45 a.m. to go to the beach with friends. She returned home at 6:30 p.m. For how long was Lizzie away with her friends?

Keywords: ? ?

1. Try It Yourself. (pages 42–43)

Question: How many hours and minutes was Lizzie away?

Keywords: left, returned, long

Facts: Lizzie left at 7:45 a.m. She returned at 6:30 p.m.

Strategy: Make a Table

Solve:

Time	Time	Elapsed Time
7:45 a.m.	8:00 a.m.	15 min
8:00 a.m.	6:00 p.m.	10 h
6:00 p.m.	6:30 p.m.	30 min

15 min + 10 h + 30 min = 10 h 45 min

Lizzie was away for 10 hours 45 minutes.

Write/Explain: I **Made a Table**. I found the elapsed times from 7:45 a.m. to 8 a.m., from 8 a.m. to 6 p.m., and from 6 p.m. to 6:30 p.m. I added the elapsed times to find that Lizzie was away for 10 hours 45 minutes.

2. Buddy Tries It. (pages 43–44)

Score the Answer: I would give Buddy a **2**. He wrote the question asked, the keywords, and the facts, named and correctly used a strategy, and tried to answer the question that was asked. He did not know how to find elapsed time since he did not use a.m. and p.m. and counted back from 7:45 a.m. instead of counting to 6:30 p.m. He did label his work and showed all his steps.

Make it a 4! Rewrite.

Time	Time	Elapsed Time
7:45 a.m.	8:00 a.m.	15 min
8:00 a.m.	6:00 p.m.	10 h
6:00 p.m.	6:30 p.m.	30 min

15 min + 10 h + 30 min = 10 h 45 min

Lizzie was away for 10 hours 45 minutes.

I **Made a Table**. I found the elapsed times from 7:45 a.m. to 8 a.m., from 8 a.m. to 6 p.m., and from 6 p.m. to 6:30 p.m. I added the elapsed times to find that Lizzie was away for 10 hours 45 minutes.

Quiz Problems

Here are some problems for you to try. Keep your **rubric** handy while you solve the problem. Let's see if you can score a **4**.

1. John's uniform number is 25 more than Manny's. Manny's uniform number is 13 less than Terry's. Terry's uniform number is twice Umberto's number. Umberto wears number 14 on his uniform. What is John's uniform number?

2. A movie theater sells children's tickets for $7 and adult tickets for $9. The theater sold 120 tickets and earned $1,000 last night. How many of each ticket did the theater sell?

3. The bill at O'Dell's came to $120. Mr. McCarthy left the waitress a 15% tip. How much money did Mr. McCarthy pay all together?

4. Amanda and Brittany sold $150 worth of tickets all together for the raffle sale. Tickets cost $2 each. If Brittany sold 40 percent of the tickets, how many tickets did Amanda sell?

5. Hot dogs come in packages of 10. Hot-dog buns come in packages of 8. How many packages of each do you need to buy to have the same number of hot dogs and hot dog buns?

6. There are 47 students in the chorus. They are going to perform at a senior citizens' home. The students will be driven there by some of the parents. If each parent's car can hold 5 students, how many cars are needed?

7. Michele's apartment number is the greatest prime number less than 100. What is her apartment number?

6. Algebra

In this chapter, we are going to look at some basic ideas of algebra. Algebra is about discovering **patterns** and **finding unknowns**. It makes you think about **number relationships** and **equations**, such as what is equal to what. You will also focus on your thoughts that lead to your best guess when solving problems.

In algebra, we use **symbols** like a box or a letter to take the place of a number. The modeled problem here will show how this is done. Let's solve it together to show what a score of **4**, model answer might look like. Then we can score it using a **rubric.**

Modeled Problem

Dawn reads a thermometer to find the temperature is 20°**C**. What is the temperature in °**F**?

Keywords: °C, °F

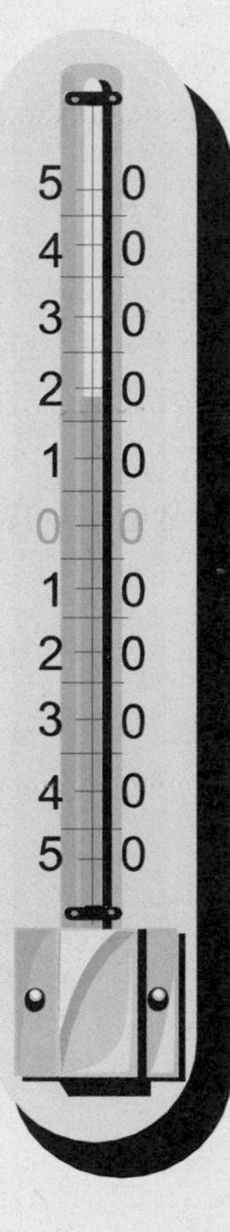

6. Algebra

1. Read and Think

What is the **question**?

- **What is the temperature in °F if it is 20°C?**

What are the **keywords**?

- **°C, °F**

What are the **facts**?

- **The temperature is 20°C.**

2. Select a Strategy

To solve this problem, we are going to **Use a Formula.**

3. Solve

The formula for converting °C to °F is $F = 1.8C + 32$.

Substitute 20 for C to get $F = 1.8 \times 20 + 32$

$$F = 36 + 32$$

$$F = 68$$

The temperature is 68°F.

4. Write/Explain

We used a formula to find the temperature in °F when the temperature is given in °C. We substituted 20 for C in the formula and then solved the equation. The temperature is 68°F.

5. Reflect

Let's review our work and answer.

- Did we show that we knew what the problem asked? **Yes. We answered the question that was asked.**
- Did we know what the keywords were? **Yes.**
- Did we show that we knew what facts were given? **Yes.**
- Did we name and use the correct strategy? **Yes. We used the formula for converting temperatures in °C to °F.**
- Was our math correct? **Yes. We checked it. It was correct.**
- Did we label our work? **Yes.**
- Was our answer correct? **Yes.**
- Were all of our steps included? **Yes.**
- Did we explain why we chose the strategy and how it was used? **Yes.**
- Did we write a good, clear explanation of our work? **Yes.**

Score

This solution would receive a **4** on our rubric.

On the following pages are some **Guided Open-Ended Math Problems**.

For each problem there are **four parts**. In the **first part**, you will solve the problem with guided help. In the **second part**, you will score and correct a solution with guided help. The **third part** shows one solution that scores a perfect **4**. This solution may or may not differ from your way. The **fourth part** has answers to the **first** and **second parts** so you can check your work.

Guided Problem #1

Mike collects baseball cards. He had 15 complete sets of cards when he bought c more complete sets. He now has 31 complete sets. How many complete sets of baseball cards did he buy?

Keywords: ? ?

1. Try It Yourself.

Answer the questions below to get a score of **4**.

What **question** are you being asked?

What are the **keywords**?

What are the **facts** you need to solve the problem?

What **strategy** can you use to solve the problem?

Hint

Possible answers include **Write an Equation**, **Act It Out**, and **Draw a Picture**.

Solve the problem.

Write/Explain what you did to solve the problem.

Reflect. Review and improve your work.

2. Roy Tries It.

Roy's Paper

Question: How many complete sets does Mike have now?

Facts: Mike had 15 complete sets of baseball cards.

He bought *c* more complete sets.

He now has 31 complete sets.

Strategy: Write a Number Sentence

Solve:

31 + 15 = 46

Write/Explain: I Wrote a Number Sentence. I added the number of complete sets that Mike had to the number that he has now. He has 46 complete sets.

Score the Answer.

According to the rubric, from **1** to **3**, what score would you give Roy? Explain why you gave that score.

Make it a 4! Rewrite.

3. Jessica Tries It.

Remember there is often more than one way to solve a problem. Here is how Jessica solved this problem.

Jessica's Paper

Question: How many complete sets did Mike buy?
Keywords: more, many
Facts: Mike had 15 complete sets of baseball cards.
He bought c more complete sets.
He now has 31 complete sets.
Strategy: I can use the Act It Out strategy.
Solve: I put 31 counters in a cup.
I took 15 counters out of the cup.
I counted the number of counters still in the cup.
There were 16 counters in the cup.
So, Mike bought 16 complete baseball card sets.

Write/Explain: I used the Act It Out strategy. I used a counter to represent each of the complete sets that Mike has now. I put the counters in a cup and I took one out for each complete set that Mike already had. The number of counters left in the cup, 16, is the number of complete sets that he bought.

Score: Jessica's solution would earn a **4** on a test.
She identified the question that was asked, the keywords, and the facts. She picked a good strategy and used it correctly. She clearly explained the steps taken to solve the problem. She labeled her answer.

4. Answers to Parts 1 and 2.

Guided Problem #1

Mike collects baseball cards. He had 15 complete sets of cards when he bought c more complete sets. He now has 31 complete sets. How many complete sets of baseball cards did he buy?

Keywords: ? ?

1. Try It Yourself. (page 52)

Question: How many complete sets did Mike buy?

Keywords: more, many

Facts: Mike has 15 complete sets of baseball cards.
He buys c more complete sets.
He now has 31 complete sets.

Strategy: Write an Equation or Act It Out

Solve: I can **Write and Solve an Equation** or **Act It Out**

$15 + c = 31$

$15 - 15 + c = 31 - 15$

$c = 16$

To **Act It Out**, I use 31 playing cards, each card representing 1 complete set. Then I take out 15 cards. There are 16 cards left.

Mike bought 16 complete sets.

Write/Explain: I **Wrote an Equation:** $15 + c = 31$. I subtracted 15 from both sides of the equation to isolate the variable. I found that $c = 16$, so Mike bought 16 complete sets of baseball cards.

I could also use the **Act It Out** strategy. I used 31 playing cards to represent the 31 complete sets. I took out 15 cards to show the number of sets Mike already has. The remaining number of sets, 16, represents the number of sets Mike bought.

2. Roy Tries It. (page 53)

Score the Answer: I would give Roy a **2**. He gave the keywords and listed the facts. His math was right. He chose and correctly used a strategy. But Roy did not write the correct question nor did he correctly answer the question that was asked. Roy added $31 + 15$ when he should have subtracted $31 - 15$.

Make it a 4! Rewrite.

Question: How many complete sets did Mike buy?

$15 + c = 31$

$15 - 15 + c = 31 - 15$

$c - 16$

Mike bought 16 complete sets.

Guided Problem #2

Mimi is working with this function table.

x	y
2	5
5	11
8	17
12	25

What is the value of y, when x is 15?

Keywords:  

1. Try It Yourself.

Answer the questions below to get a score of **4**.

What **question** are you being asked?

____

____

____

What are the **keywords**?

____

____

____

What are the **facts** you need to solve the problem?

____

____

____

What **strategy** can you use to solve the problem?

____

____

____

Hint

Possible answers include **Write an Equation** and **Look For a Pattern**.

Solve the problem.

____

____

____

Write/Explain what you did to solve the problem.

____

____

____

____

Reflect. Review and improve your work.

____

____

____

____

2. Erica Tries It.

Erica's Paper

Question: What is y when $x = 15$?

$y = 31$

Write/Explain: I found that $y = 31$ when $x = 15$.

Use the rubric on *page 13* to score this work.

Score the Answer.

According to the rubric, from **1** to **3** what score would you give Erica? Explain why you gave that score.

Make it a 4! Rewrite.

3. Fred Tries It.

Remember there is often more than one way to solve a problem. Here is how Fred solved this problem.

Fred's Paper

Question: What is y when $x = 15$?

Keywords: function, value

Facts: 2 becomes 5, 5 becomes 11, 8 becomes 17, and 12 becomes 25

Strategy: I Drew a Graph.

Solve:

$y = 31$ when $x = 15$

Write/Explain: I Drew a Graph to show the points that were given. I drew a line to connect the points. I extended the graph to find what y would be when $x = 15$. I found that $y = 31$ when $x = 15$.

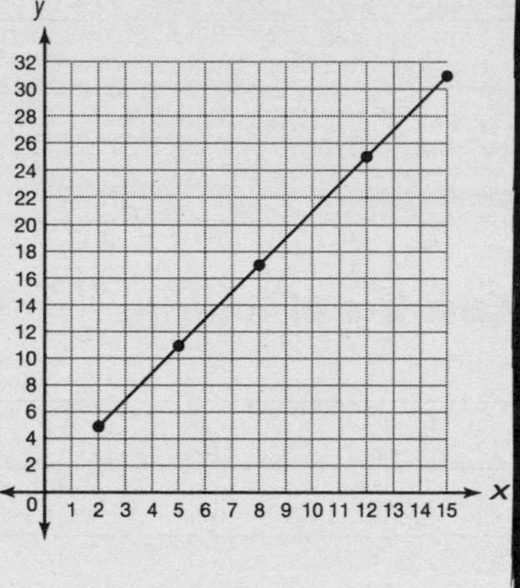

Score: Fred's solution would earn a **4** on our rubric. He identified the question that was asked, the keywords, and the facts, and picked a good strategy and used it correctly. Then he clearly explained the steps he took to solve the problem and labeled his answer.

4. Answers to Parts 1 and 2.

Guided Problem #2

Mimi is working with this function table.

x	y
2	5
5	11
8	17
12	25

What is the value of y, when x is 15?

Keywords: ? ?

1. Try It Yourself. **(page 56)**

Question: What is y, when $x = 15$?

Keywords: function, value

Facts: 2 becomes 5, 5 becomes 11, 8 becomes 17, and 12 becomes 25

Strategies: Logical Thinking and Write an Equation or Look For a Pattern

Solve: Each value of y is 1 more than double x. So, $y = 2x + 1$.

I can substitute 15 for x.

$y = 2 \times 15 + 1$

The value of $y = 31$

Write/Explain: I used **Logical Thinking** to find the rule of the function table. Since every value of y is 1 more than 2 times x, the rule of the function is $y = 2x + 1$. I substituted 15 for x and found that $y = 31$ when $x = 15$.

I can also **Look For a Pattern**. The difference between the values of y is 6 when x is 2 and when x is 5. The same applies when x is 5 and when x is 8. Since 15 is also 3 away from 12, the difference between the y values will be 6. So I added $6 + 25 = 31$ to get the value of y when x is 15.

2. Erica Tries It. (page 57)

Score the Answer: Erica would get a **1**, although she wrote the question asked and found the correct answer. She did not give the keywords or the facts. She did not name her strategy. It is impossible to tell if her mathematics is correct because she didn't show any of her steps. She did not explain what she did.

Make it a 4! Rewrite.

Keywords: function, value

Facts: 2 becomes 5, 5 becomes 11, 8 becomes 17, and 12 becomes 25

Strategies: Logical Thinking and Write an Equation.

Solve: Each value of y is 1 more than double x. So, $y = 2x + 1$.

I can substitute 15 for x.

$y = 2 \times 15 + 1$

The value of $y = 31$.

Guided Problem #3

There are 48 members in the drama club. There are three times as many girls as there are boys. Four members are also in the chess club. How many boys are in the club? How many girls are in the club?

Keywords: ? ?

1. Try It Yourself.

Answer the questions below to get a score of **4**.

What **questions** are you being asked?

What are the **keywords**?

What are the **facts** you need to solve the problem?

What **strategy** can you use to solve the problem?

Hint

Possible answers include **Guess and Test**, **Write an Equation**, and **Draw a Picture**.

Solve the problem.

Write/Explain what you did to solve the problem.

Reflect. Review and improve your work.

2. Sung Tries It.

Sung's Paper

Questions: How many girls are in the drama club? How many boys are in the drama club?
Keywords: times as many
Facts: There are 48 members of the drama club. There are 3 times as many girls as boys. Four members are also in the chess club.
Strategy: Wrote Number Sentences
Solve: There are $\frac{2}{3}$ girls: $\frac{2}{3} \times 48 = 32$.
There are $\frac{1}{3}$ boys: $\frac{1}{3} \times 48 = 16$.
There are 32 girls and 16 boys in the club.
Write/Explain: I Wrote Number Sentences to find how many boys and girls there are. I multiplied the number of members by $\frac{2}{3}$ to find the number of girls. Then I multiplied the number of members by $\frac{1}{3}$ to find the number of boys. I determined there are 32 girls and 16 boys in the drama club.

Score the Answer.

According to the rubric, from **1** to **3** what score would you give Sung? Explain why you gave that score.

Make it a 4! Rewrite.

3. Natalie Tries It.

Remember there is often more than one way to solve a problem. Here is how Natalie solved this problem.

Natalie's Paper

Questions: How many girls are in the club?
How many boys are in the club?
Keywords: times as many
Facts: There are 48 members of the club.
There are 3 times as many girls as boys.
Strategy: Draw a Picture
Solve: I made a drawing. There are 3 girls for every boy.

G G G B G G G B G G G B G G G B
G G G B G G G B G G G B G G G B
G G G B G G G B G G G B G G G B

There are 36 girls and 12 boys.

Write/Explain: I Drew a Picture to represent the situation. There are 3 girls for every boy, so I wrote 48 letters with 3 G's (girl) for every B (boy). I then counted the G's (36) and B's (12). There are 36 girls and 12 boys, which works since $36 + 12 = 48$ and $\frac{36}{12} = 3$.

Score: Natalie would get a **4** on our rubric. She showed she understood the problem, she knew the questions asked, Keywords, and facts. She named and also knew how to correctly use the strategy. She answered the question, explained her solution well, and labeled her answer. Her solution to the problem was perfect.

4. Answers to Parts 1 and 2.

Guided Problem #3

There are 48 members in the drama club. There are three times as many girls as there are boys. Four members are also in the chess club. How many boys are in the club? How many girls are in the club?

Keywords: ? ?

1. Try It Yourself. (page 60)

Questions: How many girls are in the club? How many boys are in the club?

Keywords: times as many

Facts: There are 48 members of the club. There are 3 times as many girls as boys.

Strategies: Logical Thinking and Write an Equation

Solve: Since there are 3 girls for every boy, $\frac{3}{4}$ of the members are girls and $\frac{1}{4}$ are boys.

$\frac{3}{4} \times 48 = 36$

$\frac{1}{4} \times 48 = 12$

$36 + 12 = 48$

$\frac{36}{12} = 3$

Write/Explain: Since there are 3 girls for every boy, $\frac{3}{4}$ of the members are girls and $\frac{1}{4}$ of the members are boys. I multiplied $\frac{3}{4} \times 48$ to find the number of girls and $\frac{1}{4} \times 48$ to find the number of boys. The products added to 48 and the number of girls divided by the number of boys equals 3, so my math checks out.

2. Sung Tries It. (page 61)

Score the Answer: Sung would get a **2** because he answered the question incorrectly. He knew what was asked, gave the keywords, and knew what strategy to use and how to use it. He labeled his work, but his answer is incorrect. He also wrote the extraneous information as part of the facts.

Make it a 4! Rewrite.

Take out "Four members are also in the chess club" from his facts list. Since there are 3 girls for every boy, $\frac{3}{4}$ of the members are girls and $\frac{1}{4}$ are boys.

$\frac{3}{4} \times 48 = 36$

$\frac{1}{4} \times 48 = 12$

There are 36 girls and 12 boys.

$36 + 12 = 48$

$\frac{36}{12} = 3$

The number of girls plus the number of boys equals 48 and the number of girls divided by the number of boys equals 3, so it checks out.

Quiz Problems

Here are some problems for you to try. Keep your **rubric** handy while you solve the problem. Let's see if you can score a **4**.

1. What is the seventh number in this pattern?

9, 17, 25, 33, . . .

2. Kara is working with this function table.

x	*y*
1	2
4	14
7	26
9	34
12	?

What is the missing number in the function table?

3. The temperature outside has increased by d degrees Fahrenheit each hour for 3 hours. The temperature is now 72°F. If the temperature was 57°F 3 hours before, how many degrees Farenheit did the temperature increase each hour?

4. When the bus leaves school, there are 36 students on it. After the first stop, $\frac{1}{3}$ of the students depart the bus. After the second stop, $\frac{1}{2}$ of the remaining students depart the bus. After the third stop, there are only 3 students left on the bus. How many students departed the bus at the third stop?

5. A plumber charges $80 per hour and a $35 fee for making a house call. How much does she earn if she spends $4\frac{1}{2}$ hours at a house?

6. What is the value of the ones digit in 4^{21}?

7. The temperature outside is 3°C. What is the temperature in °F?

7. Geometry

Look around your home, your school, your neighborhood. What **shapes** do you see? Squares? Cylinders? Triangles? What **lines** do you see? Some that never meet? Some that intersect? Some shorter? Some longer? What **angles** do you see? Some that are right angles? Some that are not? You are looking at geometry. Geometry is part of mathematics. Geometry is part of your everyday life.

Here is a geometry problem that might be on your tests. Let's see how we can solve this problem and get a perfect score of **4.**

We will use our **rubric** to **double-check.**

Modeled Problem

Kate drew this **net** of a **three-dimensional figure.**

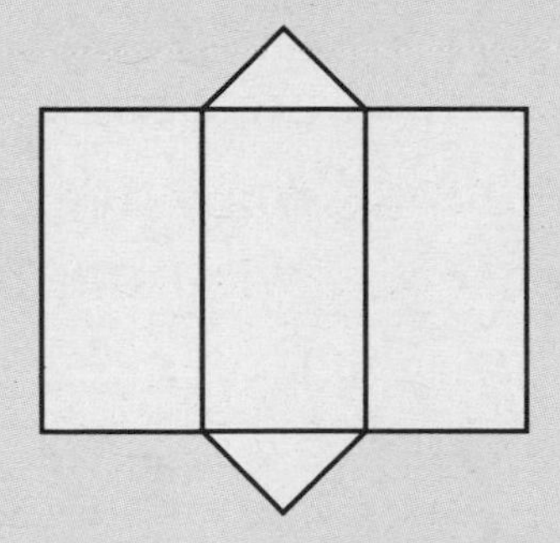

What three-dimensional figure will be formed if Kate puts the net together?

Keywords: net, three-dimensional figure

1. Read and Think

What **question** are you being asked?

- **What three-dimensional figure will be formed if the net is put together?**

What are the **keywords?**

- **net, three-dimensional figure**

What **facts** are we given?

- **The figure has 5 faces. Two are triangular and 3 are rectangular.**

2. Select a Strategy

We can use **Logical Thinking** and **Make a List**.

3. Solve

Which three-dimensional figures include triangles?

- triangular prism
- triangular pyramid
- rectangular pyramid

A triangular pyramid does not have any rectangles.

A pyramid consists mostly of triangles, so the figure is a triangular prism.

4. Write/Explain

We made a list of the three-dimensional figures that include triangles. Those shapes are triangular prism, rectangular pyramid, and triangular pyramid.
We determined that a triangular pyramid does not contain rectangles and a pyramid contains mostly triangles. Since this net shows mostly rectangles, we determined that the figure is a rectangular prism.

5. Reflect

Let review at our work and answer.

- Did we show that we knew what the question asked? **Yes. We identified the net as a triangular prism.**
- Did we know what the keywords were? **Yes.**
- Did we show that we knew what facts were given? **Yes.**
- Did we name and use the correct strategy? **Yes.**
- Was our math correct? **Yes. We checked it. It was correct.**
- Did we label our work? **Yes.**
- Was our answer correct? **Yes.**
- Were all of our steps included? **Yes.**
- Did we explain why we chose the strategy and how it was used? **Yes.**
- Did we write a good, clear explanation of our work? **Yes.**
- Was our answer correct? **Yes.**

On the following pages are some **Guided Open-Ended Math Problems**.

For each problem there are **four parts**. In the **first part**, you will solve the problem with guided help. In the **second part**, you will score and correct a solution with guided help. The **third part** shows one solution that scores a perfect **4**. This solution may or may not differ from your way. The **fourth part** has answers to the **first** and **second parts** so you can check your work.

Guided Problem #1

Roger wrote ordered pairs at (1, 3), (4, 3), and (4, 5). Identify the polygon that Roger drew. Identify your figure as specifically as you can.

Keywords: ? ?

1. Try It Yourself.

Answer the questions below to get a score of **4**.

What **question** are you being asked?

What are the **keywords?**

Hint

Possible answers include: **Draw a Picture** and **Logical Thinking**.

What are the **facts** you need to solve the problem?

What **strategy** can you use to solve the problem?

Solve the problem.

Write/Explain what you did to solve the problem.

Reflect. Review and improve your work.

2. Tony Tries It.

Tony's Paper

Question: What polygon can be made from the ordered pairs?

Keywords: ordered pairs, polygon, specifically

Facts: There are ordered pairs at (1, 3), (4, 3), (4, 5).

Strategy: Draw a Graph

Solve:

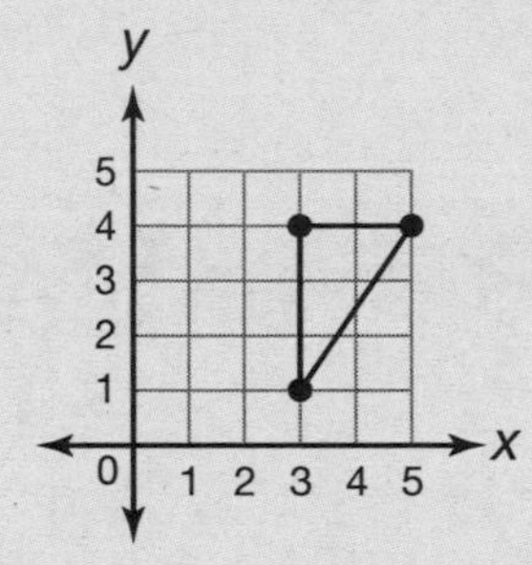

The figure has 3 points so it is a triangle.

Write/Explain: I Drew a Graph. I plotted the ordered pairs on a coordinate grid. Then I connected the points. The figure that I created has 3 sides, which is a triangle.

Score the Answer.

According to the rubric, from **1** to **3** what score would you give Tony? Explain why you gave that score.

Make it a 4! Rewrite.

3. Sean Tries It.

Remember, there is often more than one way to solve a problem. Here is how Sean solved this problem.

Sean's Paper

Question: What polygon, identified as specifically as possible, can be formed from the ordered pairs?

Keywords: ordered pairs, polygon, specifically

Facts: There are ordered pairs at (1, 3), (4, 3), and (4, 5)

Strategy: Logical Thinking

Solve:

There are three ordered pairs. A figure with 3 sides is a triangle.

The difference between the first and second coordinates is 3.

The difference between the second and third coordinates is 2.

The difference between the first and third coordinates 3 to the right and 2 up, which would be greater than 3 if I used the Pythagorean Theorem to find it.

None of the sides are equal, so the figure is a scalene triangle.

Since the first two coordinate pairs share a common coordinate (y) and the second pair of coordinate pairs also shares a common coordinate (x), the figure is a right triangle.

The figure is a scalene, right triangle.

Write/Explain: I used Logical Thinking. I counted the number of points and determined the figure was a triangle. I subtracted the difference between the coordinates. The differences between the coordinate pairs were all different, so the figure is a scalene triangle. Since each of the coordinate pairs shares a coordinate with the others, the figure is a right triangle, so it is a right, scalene triangle.

Score: Sean's solution would earn a **4** on a test.
He identified the question that was asked, the keywords, and the facts, and picked and correctly used a strategy. He clearly explained the steps taken to solve the problem, and he labeled all his work.

4. Answers to Parts 1 and 2.

Guided Problem #1

Roger wrote ordered pairs at (1, 3), (4, 3), and (4, 5). Identify the polygon that Roger drew. Identify your figure as specifically as you can.

Keywords: ? ?

1. Try It Yourself. (pages 69–70)

Question: What polygon, identified as specifically as possible, can be formed from these ordered pairs?

Keywords: ordered pairs, polygon, specifically

Facts: Points are at (1, 3), (4, 3), and (4, 5).

Strategy: Draw a Graph

Solve:

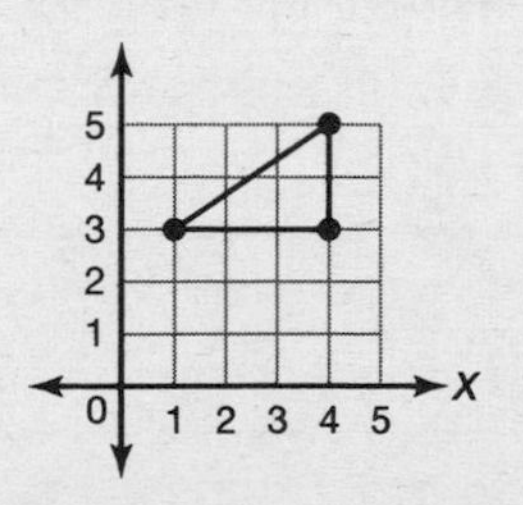

The figure is a scalene, right triangle.

Write/Explain: I **Made a Graph**. I plotted the ordered pairs on a coordinate grid. Then I connected the points. The figure that I created has 3 sides, none of which are equal. The triangle has a right angle, which makes it a scalene, right triangle.

2. Tony Tries It. (pages 70–71)

Score the Answer: I would give Tony a **2**. He knew what question was being asked, gave the keywords, and listed the facts. He also correctly picked and used a strategy. He also included all his steps and wrote a clear explanation. The answer is a triangle, but he did not identify it further. He also confused the first and second coordinates of each ordered pair, so it wasn't the correct triangle.

Make it a 4! Rewrite.

Plot the correct ordered pairs.

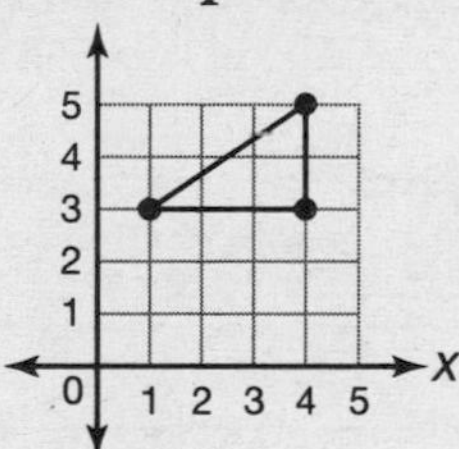

The figure has three sides, none of which are equal. Therefore, the figure is a scalene triangle. One of the angles formed is a right angle, so the figure is a scalene, right triangle.

Guided Problem #2

Does a regular hexagon tessellate? Explain why or why not.

Keywords: ? ?

1. Try It Yourself.

Answer the questions below to get a score of **4**.

What **question** are you being asked?

What are the **keywords**?

What are the **facts** you need to solve the problem?

What **strategy** can you use to solve the problem?

Hint

Possible answers include: **Write an Equation**, **Draw a Picture**, and **Logical Thinking**.

Solve the problem.

Write/Explain what you did to solve the problem.

Reflect. Review and improve your work.

2. Donna Tries It.

Donna's Paper

Question: Does a regular hexagon tessellate?

Keywords: regular hexagon, tessellate

Facts: A regular polygon has all sides equal.

Solve:

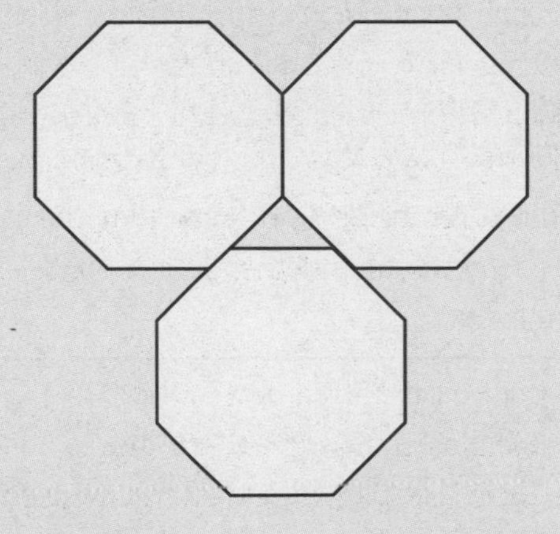

Write/Explain: A regular hexagon does not tessellate.

Score the Answer.

According to the rubric, from **1** to **3** what score would you give Dennis? Explain why you gave that score.

____

____

____

____

____

____

Make it a 4! Rewrite.

____

____

____

____

____

____

Use the rubric on *page 13* to score this work.

3. Brendan Tries It.

Remember there is often more than one way to solve a problem. Here is how Brendan solved this problem.

Brendan's Paper

Question: Does a regular hexagon tessellate?
Keywords: regular hexagon, tessellate
Facts: A regular hexagon has 6 congruent sides and angles.
Strategy: I used the Divide and Conquer strategy.
Solve: The sum of the angle measures of a hexagon equals 720°.
I can divide $\frac{720^\circ}{6} = 120^\circ$ to find the angle measure of each angle of a hexagon.
Since $\frac{360}{120} = 3$, a regular hexagon will tessellate.

Write/Explain: I used the Divide and Conquer strategy. *First,* I found the number of degrees of the interior angles of a hexagon. *Next,* I divided the sum of the angle measures by 6, since all angles are equal in a regular hexagon. The quotient was 120°. *Finally,* I divided 360 by 120 to see if there is no remainder. Since there was not a remainder, we know that the corners of the hexagon will fit together like a puzzle, so the figure tessellates.

Score: Brendan would earn a **4** on our rubric. He identified the question that was asked, the keyword, and the facts. He picked and correctly used a strategy. He explained how he used it. He clearly explained how he determined why a regular hexagon tessellates and labeled all his work.

4. Answers to Parts 1 and 2.

Guided Problem #2

Does a regular hexagon tessellate? Explain why or why not.

Keywords: ? ?

1. Try It Yourself. (page 74)

Question: Does a regular hexagon tessellate?

Keywords: regular hexagon, tessellate

Facts: A regular hexagon has all sides equal.

Strategy: Draw a Picture

Solve:

A regular hexagon tessellates since there are no overlaps or gaps.

Write/Explain: I **Drew a Picture** of 7 regular hexagons. I completely surrounded the middle regular hexagon with other regular hexagons to prove that it tessellates. There are no overlaps or gaps in the figure.

2. Donna Tries It. (page 75)

Score the Answer: I would give Donna a **2.** She understood what the question was asking, and gave the keywords and the facts. She started to use a strategy that showed she knew what a tessellation is. However, she confused an octagon with a hexagon, which gave her an incorrect answer. She included all her steps, but she didn't write out her complete strategy.

Make it a 4! Rewrite.

A regular hexagon tessellates since there are no overlaps or gaps.

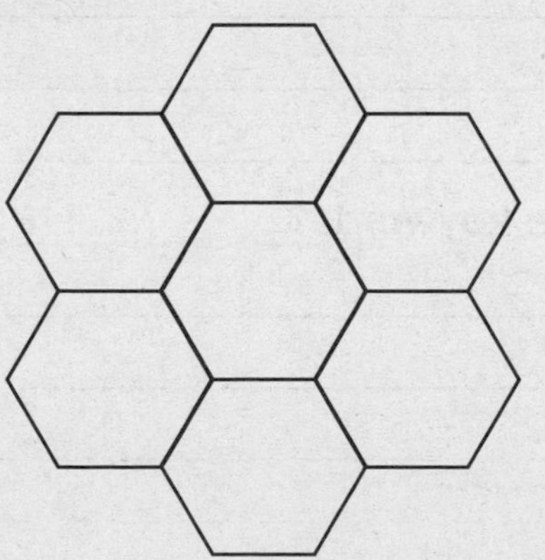

Guided Problem #3

1st Street and 2nd Street never meet and remain the same distance apart. 3rd Avenue crosses both streets at a right angle. 3rd Avenue and 4th Avenue never meet and remain the same distance apart. 4th Avenue crosses both 1st Street and 2nd Street.

Which streets form parallel lines? Which streets form perpendicular lines?

Keywords: ? ?

1. Try It Yourself.

What **question** are you being asked?

____

____

____

What are the **keywords**?

____

____

____

What are the **facts** you need to solve the problem?

____

____

____

What **strategy** can you use to solve the problem?

____

____

____

Hint

Possible answers include: **Draw a Picture**, **Make an Organized List**, and **Logical Thinking**.

Solve the problem.

____

____

____

Write/Explain what you did to solve the problem.

____

____

____

Reflect. Review and improve your work.

____

____

____

____

2. Jenna Tries It.

Jenna's Paper

Questions: Which streets are parallel? Which streets are perpendicular?

Keywords: crosses, right angle, form, parallel lines, perpendicular lines

Facts: 1st St. and 2nd St. never meet and remain the same distance apart.

3rd Ave. crosses both streets at a right angle.

3rd Ave. and 4th Ave. never meet and remain the same distance apart.

4th Ave. crosses both 1st St. and 2nd St.

Strategy: I Made an Organized List.

Solve:

Street 1	Street 2	Relationship	Type of Lines Formed
1st. St.	2nd St.	Never meet	Parallel
1st. St.	3rd Ave.	Meet at right angle	Perpendicular
3rd. Ave.	4th Ave.	Never meet	Parallel
1st. St	4th Ave.	Meet at right angle	Perpendicular

Write/Explain: I Made an Organized List to show the lines that can be formed by the streets. I know that lines that never meet and remain the same distance apart are parallel. Lines that meet at right angles are perpendicular. 1st St. is parallel to 2nd St., and 3rd Ave. is parallel to 4th Ave. 1st St. and 3rd Ave., and 1st St. and 4th Ave. are perpendicular.

Score the Answer.

According to the rubric, from **1** to **3** what score would you give Jenna? Explain why you gave that score.

Make it a 4! Rewrite.

3. Jordan Tries It.

Remember, there is often more than one way to solve a problem. Here is how Jordan solved this problem.

Jordan's Paper

Questions: Which streets form parallel lines?
Which streets form perpendicular lines?
Keywords: crosses, right angle, form, parallel lines, perpendicular lines
Facts: 1st St. and 2nd St. never meet and remain the same distance apart.
3rd Ave. crosses both streets at a right angle.
3rd Ave. and 4th Ave. never meet and remain the same distance apart.
4th Ave. crosses both 1st St. and 2nd St.
Strategy: Logical Thinking
Solve: The definition for parallel lines is two lines that never meet and remain the same distance apart. 1st St. and 2nd St. are parallel. 3rd Ave. and 4th Ave. are parallel.

The definition for perpendicular lines is two lines that cross at right angles. 3rd Ave. and 1st St. are perpendicular, as are 3rd Ave. and 2nd St. Because 4th Ave. and 3rd Ave. are parallel, 4th Ave. is also perpendicular to 1st St. and 2nd St.
Write/Explain: I used Logical Thinking. I gave the definitions for parallel lines and perpendicular lines. I used the definitions to describe the streets that are parallel and the streets that are perpendicular.

Score: Jordan would get a **4** on our rubric. She knew the questions that were asked and the keywords and the facts. She picked and correctly used an excellent strategy to find the correct answers. Her explanation was detailed, and she labeled all her work.

4. Answers to Parts 1 and 2.

Guided Problem #3

1st Street and 2nd Street never meet and remain the same distance apart. 3rd Avenue crosses both streets at a right angle.

3rd Avenue and 4th Avenue never meet and remain the same distance apart. 4th Avenue crosses both 1st Street and 2nd Street.

Which streets form parallel lines? Which streets form perpendicular lines?

Keywords: ? ?

1. Try It Yourself. (page 78)

Questions: Which streets form parallel lines? Which streets form perpendicular lines?

Keywords: crosses, right angle, form, parallel lines, perpendicular lines

Facts: 1st Street and 2nd Street never meet and remain the same distance apart.

3rd Avenue crosses both streets at a right angle.

3rd Avenue and 4th Avenue never meet and remain the same distance apart. 4th Avenue crosses both 1st Street and 2nd Street.

Strategies: Draw a Picture and Make an Organized List

Solve:

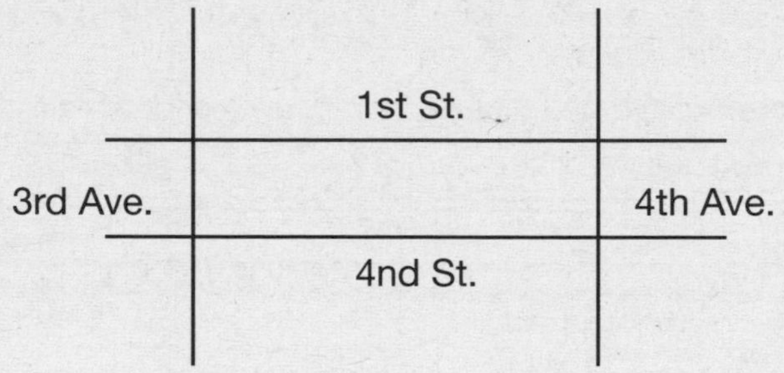

1st St.

3rd Ave.

4th Ave.

4nd St.

Parallel: 1st St. and 2nd St., 3rd Ave. and 4th Ave.

Perpendicular: 1st St. and 3rd Ave., 1st St. and 4th Ave., 2nd St. and 3rd Ave., 2nd St. and 4th Ave.

Write/Explain: I **Drew a Picture**. I listed the street and avenue names as they were listed in the question. Then I **Made an Organized List** to help me determine that 1st St. and 2nd St. were parallel as were 3rd Ave. and 4th Ave. 1st St. is perpendicular to 3rd Ave. and 4th Ave., and 2nd St. is perpendicular to 3rd Ave. and 4th Ave.

2. Jenna Tries It. (page 80)

Score the Answer: I would give Jenna a **3**. She gave the questions that were asked, the keywords, and the facts. She picked and correctly used a strategy. Most of her answer was correct, but she did not completely answer the question. She failed to notice that 2nd St. and 3rd Ave. were also perpendicular, as were 2nd St. and 4th Ave. She labeled her work. She included all her steps, but her steps were not complete enough to solve the problem. She wrote a clear explanation of her work.

Make it a 4! Rewrite.

All of the information that Jenna gave is correct. Continue the table to show that 2nd St. and 3rd Ave. are perpendicular as are 2nd St. and 4th Ave.

Quiz Problems

Here are some problems for you to try. Keep your **rubric** handy while you solve the problem. Let's see if you can score a **4.**

1. Wilson drew a pair of parallel lines. He then drew a line that is perpendicular to both parallel lines. How many right angles did Wilson form?

2. Carrie has drawn a triangle with a base of 3 inches and a height of 2 inches. Marcia has drawn a triangle with a base of 5 inches and a height of 3 inches. Are the triangles congruent, similar but not congruent, or neither?

3. Evan has plotted the ordered pair (1, 2) on a coordinate grid. He then plots the ordered pair (3, 4) on the grid. What ordered pairs does he need to plot to draw a square?

4. A quadrilateral has angle measures of 120°, 130°, and 60°. What is the measure of the fourth angle?

5. Two of the angle measures of an isosceles triangle are each 60°. What is the measure of the third angle? What else can you say about this triangle?

6. Allen plotted a square at ordered pairs (1, 2), (4, 5), (1, 5), and (4, 2). He then reflected the square across the x-axis. What are the ordered pairs for the reflected image of the square?

7. What is the greatest number of lines of symmetry that a rhombus can have? Explain your answer.

8. Measurement

There are all kinds of measurements on tests. You have to know about **length**, which includes **inches, feet, yards, centimeters,** and **meters.** You have to know about **capacity**, such as **cups, pints, quarts, gallons, milliliters,** and **liters.** You have to know about units like **ounces, pounds, grams,** and **kilograms** that are used to measure **weight** and **mass. Money, time,** and **temperature** are also kinds of measurement.

Let's look at a problem in measurement that you might have to solve on a test. Let's see how to solve it. Remember, we want to get a perfect score of **4** on our scoring **rubric**.

Modeled Problem

A playground being built is going to be **rectangular** and have a **length** of 120 feet and a **width** of 60 feet. The blacktop is about to be tarred. How **many square yards** of tar are needed?

Keywords: rectangular, length, width, many, square yards

1. Read and Think

What **question** are we asked?

- **What is the area of the playground in square yards?**

What are the **keywords?**

- **rectangular**
- **length**
- **width**
- **many**
- **square yards**

What **facts** are we given?

- **The playground has a length of 120 feet.**
- **It has a width of 60 feet.**

2. Select a Strategy

This problem has two parts. We can use the **Divide and Conquer** strategy.

The *first* part is converting feet to yards. The *second* part is finding the area.

3. Solve

1 yard = 3 feet, so divide the length and the width of the playground by 3 to find the number of yards.

$\frac{120}{3} = 40$, so the playground has a length of 40 yards.

$\frac{60}{3} = 20$, so the playground has a width of 20 yards.

To find the area of a rectangle, multiply the length times the width.

$A = lw$

$A = 40 \text{ yd} \times 20 \text{ yd}$

$A = 800 \text{ yd}^2$

The area of the playground will be 800 yd^2, so 800 yd^2 of tar is needed.

4. Write/Explain

We used the **Divide and Conquer** strategy to *find* the area of the playground in square yards. *First,* we converted the measures from feet to yards. *Then,* we used the formula for area to find the area of the playground. Since the playground is being blacktopped, it is necessary to find the area to know how much tar is needed.

5. Reflect

Let's review our work and answer.

- Did we show that we knew what the problem asked? **Yes.**
- Did we know what the keywords were? **Yes.**
- Did we show that we knew what facts were given? **Yes.**
- Did we name and use the correct strategy? **Yes.**
- Was our math correct? **Yes. We checked it. It was correct.**
- Did we label our work? **Yes.**
- Was our answer correct? **Yes. The area is 800 yd^2.**
- Were all of our steps included? **Yes.**
- Did we explain why we chose the strategy and how it was used? **Yes.**
- Did we write a good, clear explanation of our work? **Yes.**

Score

This solution would earn a **4** on our rubric. **Divide and Conquer** is a good strategy to use when a problem has more than one part.

On the following pages are some **Guided Open-Ended Math Problems.**

For each problem there are **four parts**. In the **first part**, you will solve the problem with guided help. In the **second part**, you will score and correct a solution with guided help. The **third part** shows one solution that scores a perfect **4**. This solution may or may not differ from your way. The **fourth part** has answers to the **first** and **second parts** so you can check your work.

Guided Problem #1

Phil is painting the walls and the ceiling of his bedroom. The room is 15 feet by 12 feet. The ceiling is 8 feet high. He has two windows that each measure 4 feet by 5 feet. He will use 2 coats of paint. If each gallon of paint covers 300 ft^2, how many gallons of paint does he need to buy?

Keywords: ? ?

(You will try exercises for this question on the following pages.)

8. Measurement

1. Try It Yourself.

nswer the questions below to get a score of **4**.

What **question** are you being asked?

What are the **keywords**?

What are the **facts** you need to solve the problem?

What **strategy** can you use to solve the problem?

Hint

Possible answers include: **Divide and Conquer** and **Draw a Picture**.

Solve the problem.

Write/Explain what you did to solve the problem.

Reflect. Review and improve your work.

2. Trish Tries It.

Trish's Paper

Questions: What is the surface area that Phil will paint? How many gallons of paint does Phil need?

Keywords: each, many, gallon

Facts: The room is 15 feet by 12 feet.

The ceiling is 8 feet high.

There are two windows that each measure 4 feet by 5 feet.

He will use 2 coats of paint.

A gallon of paint covers 300 ft^2.

Strategy: Divide and Conquer

Solve: Wall 1: $15 \text{ ft} \times 8 \text{ ft} = 120 \text{ ft}^2$

Wall 2: $12 \text{ ft} \times 8 \text{ ft} = 96 \text{ ft}^2$

Ceiling: $15 \text{ ft} \times 12 \text{ ft} = 180 \text{ ft}^2$

Window 1: $4 \text{ ft} \times 5 \text{ ft} = 20 \text{ ft}^2$

Window 2: $4 \text{ ft} \times 5 \text{ ft} = 20 \text{ ft}^2$

$120 \text{ ft}^2 + 96 \text{ ft}^2 + 180 \text{ ft}^2 + 20 \text{ ft}^2 + 20 \text{ ft}^2 = 436 \text{ ft}^2$

$436 \text{ ft}^2 \times 2 = 872 \text{ ft}^2$

$\frac{872 \text{ ft}^2}{300 \text{ ft}^2} = 2 \text{ R}272$

Phil needs 3 gallons of paint.

Write/Explain: I used the Divide and Conquer strategy. I found the area of the walls, the ceiling, and the windows. Then I added the areas and multiplied by 2 since Phil is using 2 coats. I divided the sum of the areas by 300 to find how many gallons of paint are needed. Since there was a remainder, I added 1 to the quotient.

Score the Answer.

According to the rubric, from **1** to **3** what score would you give Trish? Explain why you gave that score.

____

____

____

____

Make it a 4! Rewrite.

____

____

____

____

____

____

____

3. Francesca Tries It.

Remember there is often more than one way to solve a problem. Here is how Francesca solved this problem.

Francesca's Paper

Questions: What is the surface area that Phil will paint? How many gallons of paint does Phil need?

Keywords: each, many, gallon

Facts: The room is 15 feet by 12 feet.
The ceiling is 8 feet high.
There are 2 windows that each measure 4 feet by 5 feet.
He will use 2 coats of paint.
A gallon of paint covers 300 ft^2.

Strategies: I Wrote Number Sentences and Used Divide and Conquer.

Solve: Wall: $2 \times 2 \times 15 \times 8 = 480\ ft^2$
Wall: $2 \times 2 \times 12 \times 8 = 384\ ft^2$
Ceiling: $2 \times 15 \times 12 = 360\ ft^2$
Total: $480 + 384 + 360 = 1{,}224\ ft^2$
Windows: $2 \times 2 \times 4 \times 5 = 80\ ft^2$
$1{,}224\ ft^2 - 80\ ft^2 = 1{,}144\ ft^2$
$1{,}144\ ft^2/300 = 3\ R244$
Phil needs 4 gallons of paint.

Write/Explain: I Wrote Number Sentences and used the Divide and Conquer strategy. I found the area of each of the longer walls, the shorter walls, the ceiling, and the windows. I multiplied each by 2 since there were 2 of each wall and by 2 again since Phil is using 2 coats of paint. I subtracted the surface area of the windows from the total surface area of the walls. I then divided the area by 300 since each coat can cover 300 ft^2. Since there was a remainder, I added 1 to the quotient.

Score: Francesca would earn a **4** on the rubric. She knew the questions that she needed to answer, gave the keywords, and listed the facts. She correctly chose and used her strategies. Her strategies helped her to keep her facts straight. She clearly explained and labeled her work. Her math was correct, as was her answer.

4. Answers to Parts 1 and 2.

Guided Problem #1

Phil is painting the walls and the ceiling of his bedroom. The room is 15 feet by 12 feet. The ceiling is 8 feet high. He has two windows that each measure 4 feet by 5 feet. He will use 2 coats of paint. If each gallon of paint covers 300 ft^2, how many gallons of paint does he need to buy?

Keywords: ? ?

1. Try It Yourself. (pages 89–90)

Questions: What is the surface area that Phil will paint?
How many gallons of paint does Phil need?

Keywords: each, many, gallon

Facts: The room is 15 feet by 12 feet.
The ceiling is 8 feet high.
There are 2 windows that each measure 4 feet by 5 feet.
He will use 2 coats of paint.
A gallon of paint covers 300 ft^2.

Strategies: Make a Table and Divide and Conquer

Solve:

	Length	Width	Area in $ft.^2$	Total Area in $ft.^2$
Wall 1	15 ft.	8 ft.	120	120
Wall 2	15 ft.	8 ft.	120	240
Wall 3	12 ft.	8 ft.	96	336
Wall 4	12 ft.	8 ft.	96	432
Ceiling	15 ft.	12 ft.	180	612
Window 1	4 ft.	5 ft.	20	592
Window 2	4 ft.	5 ft.	20	572

572 $ft^2 \times 2 = 1{,}144\ ft^2$

1,144 $ft^2 \div 300 = 3$ R244

Phil needs 4 gallons of paint.

Write/Explain: I **Made a Table** and used the **Divide and Conquer** strategy. I found the area of each of the walls, the windows, and the ceiling. I kept a running total of the area and subtracted the area of the windows from the total. I multiplied the total area by 2 since Phil is using 2 coats of paint. I then divided the area by 300 since each coat can cover 300 ft^2. Since there was a remainder, I added 1 to the quotient.

2. Trish Tries It. (page 91)

Score the Answer: I would give Trish a **2**. She knew what the question was asking, gave the keywords, and listed the facts. She correctly chose and used a strategy. She labeled her work and included all her steps. And Trish wrote a clear explanation of her work. However, she did not find the area of all of the walls, which caused her to write an incorrect answer. There were two 8 by 15 walls, and two 8 by 12 walls.

Make it a 4! Rewrite.

$A = 240 + 192 + 180 - 40 = 572^2$

	Number	Length	Width	Area in ft.^{2}	Total Area in ft.^{2}
Longer walls	2	15 ft	8 ft	120	240
Shorter Walls	2	12 ft	8 ft	96	192
Ceiling		15 ft	12 ft	180	180
Windows	2	4 ft	5 ft	20	40

$572 \text{ ft}^2 \times 2 = 1{,}144 \text{ ft}^2$

$1{,}144 \text{ ft}^2 \div 300 = 3 \text{ R}244$

Phil needs 4 gallons of paint.

I **Made a Table** and used the **Divide and Conquer** strategy. I found the area of each of the walls, the windows, and the ceiling. I kept a running total of the area and subtracted the area of the windows from the total. I multiplied the total area by 2 since Phil is using 2 coats of paint. I then divided the area by 300 since each coat can cover 300 ft^2. Since there was a remainder, I added 1 to the quotient.

Guided Problem #2

The houses on Von Bargen Avenue are each 25 meters long. They are 40 meters from each other. The corner houses are at least 50 meters away from the intersecting streets. Von Bargen Avenue is 500 meters long. How many houses are on each side of Von Bargen Avenue?

Keywords: ? ?

1. Try It Yourself.

Answer the questions below to get a score of **4.**

What **question** are you being asked?

____

____

What are the **keywords?**

____

____

____

What are the **facts** you need to solve the problem?

____

____

____

What **strategy** can you use to solve the problem?

____

____

____

Hint

Possible answers include: **Draw a Picture**, **Make a Table**, and **Make a List**.

Solve the problem.

____

____

____

Write/Explain what you did to solve the problem.

____

____

____

Reflect. Review and improve your work.

____

____

____

2. Gina Tries It.

Gina's Paper

Question: How many houses are on each side of the avenue?
Keywords: each, long, at least, away, many
Facts: The houses are 25 meters long.
The houses are 40 meters from each other.
The corner houses are 50 meters away from the intersecting streets.
The street is 500 meters long.
Strategy: I Made an Organized List.
Solve: 0-25 meters: House 1
65-90 meters: House 2
130-155 meters: House 3
195-220 meters: House 4
260-285 meters: House 5
325-350 meters: House 6
390-415 meters: House 7
455-480 meters: House 8
There are 8 houses on each side of Von Bargen Ave.
Write/Explain: I Made a List. I made each house 25 meters long and there are 40 meters between the houses. I then listed each house until I reached 500 meters.

Score the Answer.

According to the rubric, from **1** to **3** what score would you give Gina? Explain why you gave that score.

Make it a 4! Rewrite.

3. Jorge Tries It.

Remember there is often more than one way to solve a problem. Here is how Jorge solved this problem.

Jorge's Paper

Question: How many houses are on each side of the avenue?

Keywords: each, long, at least, away, many

Facts: The houses are 25 meters long.

The houses are 40 meters apart from each other.

The corner houses are 50 meters away from the intersecting streets.

The street is 500 meters long.

Strategy: I Looked For a Pattern.

Solve: Since each house is 25 meters long and each house must be 40 meters away from the next house, I added 65 meters for each house starting from 50.

50: House 1
115: House 2
180: House 3
245: House 4
310: House 5
375: House 6
440: Not enough meters to have another house.

Write/Explain: I Looked For a Pattern. I started from 50 meters since the first house is at least 50 meters from the intersecting street. All of the houses are 25 meters long and 40 meters away from the next house, so I found a pattern of 65 meters from the start of one house to the start of the next. I added 65 to the location of each house until I reached 375. Since 375 + 25 + 40 = 440, the last house would have started at 440 meters, but there is not enough space for the house to be built before the next intersecting street. There are 6 houses on each side of the avenue.

Score: Jorge's solution would earn a 4 on a test. He identified the question that was asked, the keywords, and the facts. He picked and correctly used a good strategy. He explained how he used it. He clearly explained the steps taken to solve the problem and labeled his work.

4. Answers to Parts 1 and 2.

Guided Problem #2

The houses on Von Bargen Avenue are each 25 meters long. They are 40 meters from each other. The corner houses are at least 50 meters away from the intersecting streets. Von Bargen Avenue is 500 meters long. How many houses are on each side of Von Bargen Avenue?

Keywords: ? ?

1. Try It Yourself. (pages 94–95)

Question: How many houses are on each side of the avenue?

Keywords: each, long, at least, many, away

Facts: The houses are 25 meters long.
The houses are 40 meters from each other.
The corner houses are 50 meters away from the intersecting streets. The street is 500 meters long.

Strategy: Make a Table

Solve:

Distance from corner	50-75	115-140	180-205	245-270	310-335	375-400
Houses	1	2	3	4	5	6

Write/Explain: I **Made a Table**. I made sure that each of the corner properties is at least 50 meters away from the intersecting streets. The houses are 40 meters apart and each house is 25 meters long. There are 6 houses on each side of the avenue.

2. Gina's Paper. (pages 95–96)

Score the Answer: I would give Gina a **2.** She knew what question was asked, and gave the keywords and the facts. She picked and correctly used a strategy. She also labeled her work and included all her steps. However, she did not follow the instructions of the question. She started her list from 0 and found as many houses as she could until reaching 500, but she forgot that the houses were at least 50 meters from the intersecting street, so her answer was incorrect.

Make it a 4! Rewrite.

50–75 meters: House 1

115–140 meters: House 2

180–205 meters: House 3

245–270 meters: House 4

310–335 meters: House 5

375–400 meters: House 6

Strategy: Make an Organized List

Write/Explain: I made sure that each of the corner properties are at least 50 meters away from the intersecting streets. The houses are 40 meters apart and each house is 25 meters long. There are 6 houses on each side of the avenue.

Guided Problem #3

Ken brought 4 gallons of bottled water to his soccer game. Mrs. Jennings brought 20 one-quart containers of bottled water for the team. Coach Wilson brought a cooler that holds 500 fluid ounces of water. Brenda brought 3 pounds of trail mix. Who brought the most water? Order the amounts of water from least to greatest.

Keywords: 

1. Try It Yourself.

Answer the questions below to get a score of **4.**

What **questions** are you being asked?

What are the **keywords?**

What are the **facts** you need to solve the problem?

What **strategy** can you use to solve the problem?

Hint

Possible answers include: **Divide and Conquer** and **Write a Number Sentence**.

Solve the problem.

Write/Explain what you did to solve the problem.

Reflect. Review and improve your work.

2. Dana Tries It.

Dana's Paper

Question: Who brought the most water?
What is the order from least to greatest?
Keywords: gallons, quart, fluid ounces, most, least, greatest
Facts: Ken brought 4 gallons.
Mrs. Jennings brought 20 quarts.
Coach Wilson brought 500 fluid ounces.
Brenda brought 3 pounds of trail mix.
Strategy: Write Number Sentences
Solve:
$4 \times 128 = 512$
$20 \times 32 = 640$
500
Mrs. Jennings brought the most water.
Write/Explain: I Wrote Number Sentences.
I converted gallons to fluid ounces and quarts to fluid ounces. Since 640 is greater than 512 and 500,
Mrs. Jennings brought the most water.

Score the Answer.

According to the rubric, from **1** to **3** what score would you give Dana? Explain why you gave that score.

Make it a 4! Rewrite.

3. Sophie Tries It.

Remember there is often more than one way to solve a problem. Here is how Sophie solved this problem.

Sophie's Paper

Question: Who brought the most water?
What is the order from least to greatest?

Keywords: gallons, quart, fluid ounces, most, least, greatest

Facts: Ken brought 4 gallons.
Mrs. Jennings brought 20 quarts.
Coach Wilson brought 500 fluid ounces.

Strategy: I used Logical Thinking.

Solve:
1 gallon = 4 quarts, so 4 gallons < 20 quarts
1 quart = 32 fluid ounces, so 20 quarts = 640 fluid ounces, 20 quarts > 500 fluid ounces
1 gallon = 128 fluid ounces
128 × 4 = 512, so 4 gallons > 500 fluid ounces
Mrs. Jennings brought the most water.
From least to greatest the amounts are 500 fl oz, 4 gal, 20 qt.

Write/Explain: I used Logical Thinking. I converted gallons to quarts and found that 4 gallons is less than 20 quarts. I converted quarts to fluid ounces and found that 20 quarts is greater than 500 fluid ounces, so Mrs. Jennings brought the most water. I converted gallons to fluid ounces and found that 4 gallons is greater than 500 fluid ounces. I then ordered the amounts of water from least to greatest.

Score: Sophie would earn a **4** on our rubric. She identified the question that was asked, the keywords, and the facts. She picked a good strategy and used it correctly. She clearly explained and labeled her work and gave the correct answer.

4. Answers to Parts 1 and 2.

Guided Problem #3

Ken brought 4 gallons of bottled water to his soccer game. Mrs. Jennings brought 20 one-quart containers of bottled water for the team. Coach Wilson brought a cooler that holds 500 fluid ounces of water. Brenda brought 3 pounds of trail mix. Who brought the most water? Order the amounts of water from least to greatest.

Keywords: ? ?

1. Try It Yourself. (pages 99–100)

Question: Who brought the most water? What is the order from least to greatest?

Keywords: gallons, quart, fluid ounces, most, least, greatest

Facts: Ken brought 4 gallons. Mrs. Jennings brought 20 quarts. Coach Wilson brought 500 fluid ounces.

Strategies: Make a Table and Logical Thinking

Solve:

Person	Unit	Conversion to Fl Oz	Fluid Ounces
Ken	4 gal	128×4	512
Mrs. Jennings	20 qt	32×20	640
Coach Wilson	500 fl oz		500

Mrs. Jennings brought the most water. Order from least to greatest is 500 fl oz, 4 gal, 20 qt.

Write/Explain: I **Made a Table.** I converted gallons to fluid ounces by multiplying by 128 and quarts to fluid ounces by multiplying by 32 and then compared the numbers. Mrs. Jennings brought the most water. I then used logical thinking to write the amounts of water from least to greatest.

2. Dana Tries It. (pages 100–101)

Score the Answer: I would give Dana a **2**. She knew the question, and gave the keywords, but forgot to answer the second part of the question. She also wrote the extraneous information as part of the facts and did not label thc watcr amounts. Her math was correct. And she picked and used a strategy correctly.

Make it a 4! Rewrite.

4 gal $\times$ 128 fl oz = 512 fl oz

20 qt $\times$ 32 fl oz = 640 fl oz

500 fl oz

Least to greatest: 500 fl oz, 4 gal, 20 qt.

I **Wrote Number Sentences** and converted 4 gallons and 20 quarts into fluid ounces. I then compared and ordered the numbers to answer the questions.

Quiz Problems

Here are some problems for you to try. Keep your rubric handy while you solve the problem. Let's see if you can score a **4**.

1. A swimming pool is 60 feet long, 40 feet wide, and 5 feet deep. How much water can the swimming pool hold if it is completely full?

2. The rim on a basketball court stands 120 inches above the ground. When Lester extends his arms, his hand reaches 6 feet 4 inches. How high must Lester jump for him to reach the rim? Write your answer in feet and inches.

3. What is the greatest area that a rectangle with a perimeter of 20 inches can have?

4. Barry bought a rug that is circular. The diameter of the rug is 3 feet. What is the approximate area of the rug? Use 3.14 for π.

5. Sydney wants to run 3,000 meters today. She has run 1.2 kilometers so far. How much farther does she have to run to meet her goal? Write your answer in meters.

6. Mrs. Hindes's rose garden is shaped like a right triangle. The sides of her garden are 12 feet, 16 feet, and 20 feet. What is the area of the garden?

7. Mr. Martin's basement does not have windows. The room is 24 feet by 15 feet. The ceiling is 7 feet high. What is the surface area of the basement including the ceiling and floor?

9. Data Analysis and Probability

What would you do if you had a lot of **data** to analyze? How would you **make sense** of it? How would you make it worthwhile and not wasted? How would you organize it? The data should be organized in some way. You may choose to use **tables, line plots, bar graphs,** or **line graphs**. You should choose the way you think displays the data best. When you do so, you can study the data in a logical way, and use it to draw conclusions.

It is important that you be able to solve open-ended math problems that deal with **data** and **graphs.** Let's look at a modeled problem and see how it is done.

Modeled Problem

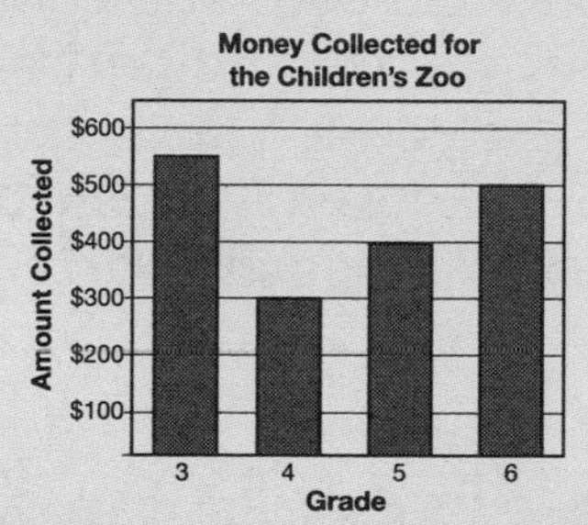

Steve is playing a game with two number cubes. If he tosses a **combined** roll of 5 or 10 he will win the game. What is the **probability** that Steve will win on this toss?

Keywords: combined, probability

1. Read and Think

What **question** are we asked?

- **What is the probability that Steve will toss a 5 or a 10?**

What are the **keywords?**

- **combined**
- **probability**

What **facts** are we given?

- **Steve is tossing 2 number cubes.**
- **Number cubes are labeled 1, 2, 3, 4, 5, 6.**

2. Select a Strategy

We will use **Draw a Picture** and **Logical Thinking** to solve the problem.

3. Solve

There are two number cubes, so we can use the Fundamental Counting Principle to find how many possible outcomes there are. There are 6 possible outcomes for the first toss and 6 possible outcomes for the second toss. Multiply $6 \times 6 = 36$. There are 36 possible outcomes. Next, make a list of how to toss a 5 or a 10.

5: 1 and 4, 2 and 3, 3 and 2, 4 and 1

10: 4 and 6, 5 and 5, and 6 and 4

There are 4 ways to roll a 5 and 3 ways to roll a 10. The probability that Steve will win on his next toss is $\frac{7}{36}$.

4. Write/Explain

We found the number of possible outcomes for tossing two number cubes by using the Fundamental Counting Principle. Then we made a list of the ways that a 5 and a 10 can be tossed with two number cubes. There were 7 different ways to toss a 5 or a 10.

The probability is $\frac{7}{36}$ that Steve win will on his next toss.

5. Reflect

Let's review our work and answer.

- Did we show that we knew what the problem asked? **Yes.**
- Did we know what the keywords were? **Yes.**
- Did we show that we knew what facts were given? **Yes.**
- Did we name and use the correct strategy? **Yes.**
- Was our math correct? **Yes. We checked it. It was correct.**
- Did we label our work **Yes.**
- Was our answer correct? **Yes.**
- Were all of our steps included? **Yes.**
- Did we explain why we chose the strategy and how it was used? **Yes.**
- Did we write a good, clear explanation of our work? **Yes.**

Let's see how some students answered other questions about handling data and information.

On the following pages are some **Guided Open-Ended Math Problems**.

For each problem there are **four parts**. In the **first part**, you will solve the problem with guided help. In the **second part**, you will score and correct a solution with guided help. The **third part** shows one solution that scores a perfect **4**. This solution may or may not differ from your way. The **fourth part** has answers to the **first** and **second parts** so you can check your work.

Guided Problem #1

The circle graph shows the results of a survey asking students to give the team a name.

Nickname Vote

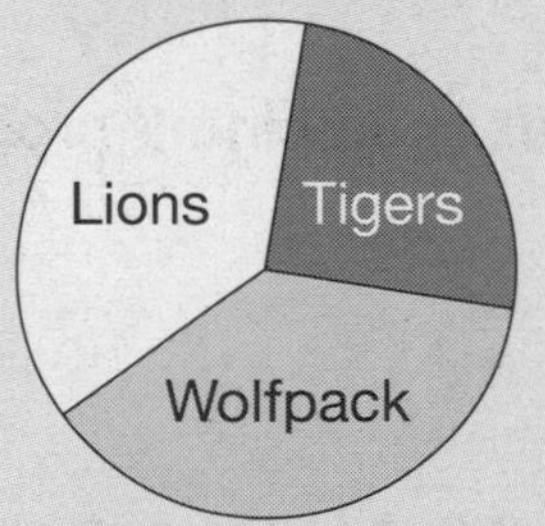

If 180 students voted, how many voted for Tigers?

Keywords: ? 

1. Try It Yourself.

Answer the questions below to get a score of **4**.

What **question** are you being asked?

What are the **keywords**?

What are the **facts** you need to solve the problem?

What **strategy** can you use to solve the problem?

Hint

Possible answers include: **Write an Equation** and **Draw a Picture**.

Solve the problem.

Write/Explain what you did to solve the problem.

Reflect. Review and improve your work.

Use the rubric on *page 13* to score this work.

2. Kate Tries It.

Kate's Paper

Question: How many of the 180 students voted for Tigers?
Keywords: circle graph, survey, many
Facts: $\frac{1}{3}$ of the students voted for Tigers.
180 students voted.
Strategy: Write a Number Sentence
Solve:

$$\frac{1}{3} \times 180 = \frac{180}{3} = 60$$

60 students voted for Tigers.

Write/Explain: I Wrote a Number Sentence.
I multiplied $\frac{1}{3}$ by the number of students to find how many students voted for Tigers.

Use the rubric on *page 13* to score this work.

Score the Answer.

According to the rubric, from **1** to **3** what score would you give Kate? Explain why you gave that score.

Make it a 4! Rewrite.

3. Trevor Tries It.

Remember there is often more than one way to solve a problem. Here is how Trevor solved this problem.

Trevor's Paper

Question: How many students voted for Tigers?

Keywords: circle graph, survey, many

Facts: $\frac{1}{4}$ of the students voted for Tigers.

180 students voted.

Strategies: I used Logical Thinking and Wrote an Equation.

Solve:

Since $\frac{1}{4}$ of the students voted for Tigers, I divided $\frac{180}{4}$.

$\frac{180}{4} = 45$

So, 45 students voted for Tigers.

Write/Explain: I used Logical Thinking and Wrote an Equation. Multiplying a unit fraction is the same as dividing by the denominator. I divided 180 by 4 to find that 45 students voted for Tigers.

Score: Trevor's solution would earn a **4** on a test. He identified the question that was asked, the keywords, and the facts. He picked a strategy and used it correctly. His logic was correct, and he wrote an equation that gave him the correct answer. He showed all of his steps and labeled his work. Trevor also clearly explained his work.

4. Answers to Parts 1 and 2.

Guided Problem #1

The circle graph shows the results of a survey asking students to give the team a name.

Nickname Vote

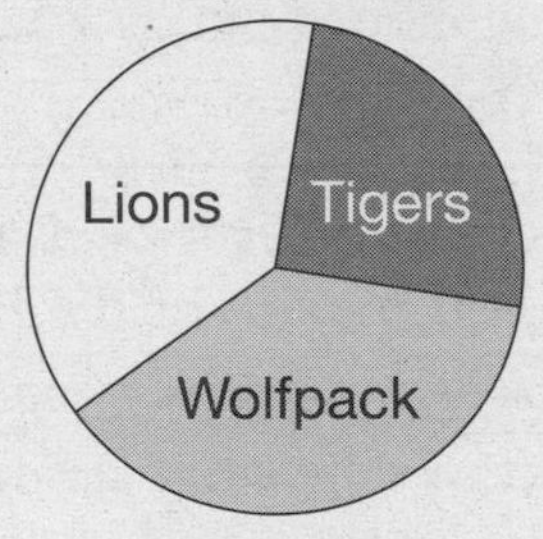

If 180 students voted, how many voted for Tigers?

Keywords:

1. Try It Yourself. (pages 108–109)

Question: How many students voted for Tigers?

Keywords: circle graph, survey, many

Facts: $\frac{1}{4}$ of the students voted for Tigers.

180 students voted.

Strategy: Write a Number Sentence

Solve: $180 \times \frac{1}{4} = 45$

Write/Explain: I interpreted the graph and **Wrote a Number Sentence**. I multiplied the number of students who voted by the fraction of those who voted for the Tigers.

2. Kate's Paper. (page 110)

Score the Answer: I would give Kate a **2**. She knew the question asked, and the keywords and listed the facts correctly. She picked a good strategy and used it correctly. She also labeled her answer, included all her steps, and wrote a clean explanation. Her math was correct, but she multiplied by the wrong fraction. She multiplied by the number of choices instead of by the fraction of students, $\frac{1}{4}$, who voted for Tigers.

Make it a 4! Rewrite.

Facts: $\frac{1}{4}$ of the students voted for Tigers.

$\frac{1}{4} \times 180 = 45$

45 students voted for Tigers.

Guided Problem #2

The ages of the family members that attended Grandma's birthday party are listed below.

45, 36, 42, 13, 15, 10, 8, 70, 42, 3

What is the median age of those who attended the party?

Keywords: ? ?

1. Try It Yourself.

Answer the questions below to get a score of **4.**

What **question** are you being asked?

What are the **keywords?**

What are the **facts** you need to solve the problem?

What **strategy** can you use to solve the problem?

Hint

Possible answers include: **Draw a Graph** and **Make an Organized List**.

Solve the problem.

Write/Explain what you did to solve the problem.

Reflect. Review and improve your work.

2. Alex Tries It.

Alex's Paper

The median is the number that occurs the most. There are two 42's, so 42 is the median.

Score the Answer.

According to the rubric, from **1** to **3** what score would you give Alex? Explain why you gave that score.

Make it a 4! Rewrite.

3. Bonnie Tries It.

Remember there is often more than one way to solve a problem. Here is how Bonnie solved this problem.

Bonnie's Paper

Question: What is the median age?

Keyword: median

Facts: Ages are 45, 36, 42, 13, 15, 10, 8, 70, 42, and 3

Strategy: I can Draw a Graph.

Solve:

Stem	Leaves
0	3 8
1	0 3 5
3	6
4	2 2 5
7	0

Key: 1 | 3 = 13

The median is the mean of the middle two numbers:
15 + 36 = 51, 51 ÷ 2 = 25.5.

The median age at the party is 25.5 years old.

Write/Explain: I made a stem-and-leaf plot to order the ages from least to greatest. Then I found the median by finding the mean of the two middle numbers. I found the mean of the two middle numbers because there is an even number of data.

Score: Bonnie's solution would earn a **4** on a test. She identified the question that was asked, the keywords, facts, picked a strategy, explained why she chose it, and used it correctly. She clearly explained the steps taken to solve the problem and labeled her answer.

4. Answers to Parts 1 and 2.

Guided Problem #2

The ages of the family members that attended Grandma's birthday party are listed below.

45, 36, 42, 13, 15, 10, 8, 70, 42, 3

What is the median age of those who attended the party?

Keywords: ? ?

1. Try It Yourself. (pages 113–114)

Question: What is the median age?

Keyword: median

Facts: The ages are 45, 36, 42, 13, 15, 10, 8, 70, 42, 3.

Strategy: Write a Number Sentence

Solve: Order the numbers from least to greatest:

3, 8, 10, 13, 15, 36, 42, 42, 45, 70

The median is the mean of the two middle numbers: 15 + 36 = 51, 51 ÷ 2 = 25.5

The median is 25.5 years old.

Write/Explain: I ordered the ages from least to greatest and then I used a **Number Sentence** to find the mean of the two middle numbers because there is an even number of data.

2. Alex Tries It. (page 114)

Score the Answer: I would give Alex a **1.** He did not write the question, the keyword, or the facts. He did not name a strategy or show how he used it. He confused the mode with the median. He did give the correct mode and gave an explanation of how he found the mode.

Make it a 4! Rewrite.

Question: What is the median age?

Keyword: median

Facts: Ages are 45, 36, 42, 13, 15, 10, 8, 70, 42, and 3.

Strategy: Write a Number Sentence

Solve:

Order the numbers from least to greatest:

3, 8, 10, 13, 15, 36, 42, 42, 45, 70

The median is the mean of the two middle numbers: 15 + 36 = 51, 51 ÷ 2 = 25.5

The median is 25.5 years old.

I listed the ages from least to greatest and then I found the mean of the two middle numbers because there is an even number of data.

Guided Problem #3

The double-bar graph shows the number of boys and girls at Evie's school who belong to clubs. Are there more boys or girls in clubs? How many more?

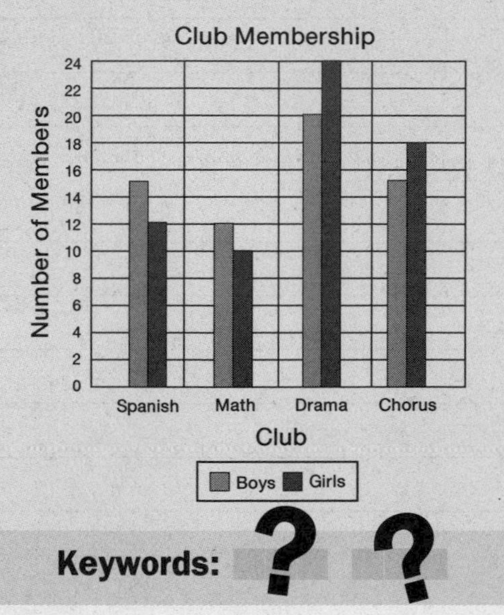

Keywords: ? ?

1. Try It Yourself.

Answer the questions below to get a score of **4**.

What **question** are you being asked?

What are the **keywords**?

What are the **facts** you need to solve the problem?

What **strategy** can you use to solve the problem?

Hint

Possible answers include: **Make a Table**, **Divide and Conquer**, and **Write an Equation**.

Solve the problem.

Write/Explain what you did to solve the problem.

Reflect. Review and improve your work.

2. Graham Tries It.

Graham's Paper

Question: Are there more boys or girls in clubs?
Keywords: double-bar graph, more
Facts: The double-bar graph shows the data.
Strategy: Divide and Conquer
Solve:
Boys: $15 + 12 + 20 + 15 = 62$
Girls: $12 + 10 + 24 + 18 = 64$

There are 126 students in the clubs.

Write/Explain: I read the graph and used the Divide and Conquer strategy. I added each of the bars to find the number of boys and then the girls. Then I added the number of boys and girls.

Use the rubric on *page 13* to score this work.

Score the Answer.

According to the rubric, from **1** to **3** what score would you give Graham? Explain why you gave that score.

____

____

____

____

____

____

Make it a 4! Rewrite.

____

____

____

____

____

____

3. Jackson Tries It.

Remember there is often more than one way to solve a problem. Here is how Jackson solved this problem.

Jackson's Paper

Questions: Are there more boys or girls in the clubs? How many more does the group that has more have than the group that has less?

Keywords: double-bar graph, more

Facts: There are 15 boys and 12 girls in the Spanish Club, 12 boys and 10 girls in the Math Club, 20 boys and 24 girls in the Drama Club, and 15 boys and 18 girls in the Chorus.

Strategy: I Made a Table.

Solve:

Boys		Girls
15	Spanish	12
12	Math	10
20	Drama	24
15	Chorus	18
62	**Total**	64

There are 2 more girls than boys in the clubs.

Write/Explain: I Made a Table from the graph. I added the number of boys in all the clubs. Then I added the number of girls in all the clubs. There were more girls than boys. Then I subtracted the number of boys from the number of girls. There were 2 more girls than boys in the clubs.

Score: Jackson would get a **4** on our rubric. He knew the question that was asked, and included the keywords and all of the facts. He chose and used an effective strategy to find the correct answer. He explained how he used it. He gave a clear explanation of what he did and labeled his answer. His work was perfect!

4. Answers to Parts 1 and 2.

Guided Problem #3

The double-bar graph shows the number of boys and girls at Evie's school who belong to clubs. Are there more boys or girls in clubs? How many more?

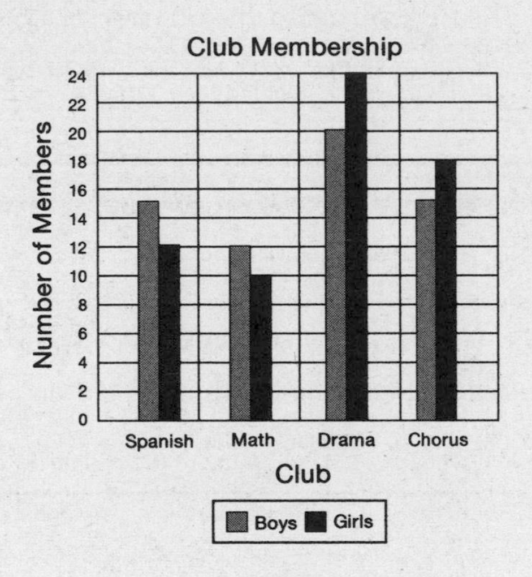

Keywords: ? ?

1. Try It Yourself. (page 117)

Questions: Are there more boys or girls in the clubs?

How many more does the group that has more have than the group that has less?

Keywords: double-bar graph, more

Facts: There are 15 boys and 12 girls in the Spanish Club, 12 boys and 10 girls in the Math Club, 20 boys and 24 girls in the Drama Club, and 15 boys and 18 girls in the Chorus.

Strategy: Write a Number Sentence

Solve: I **Wrote Number Sentences.**

Boys: $15 + 12 + 20 + 15 = 62$

Girls: $12 + 10 + 24 + 18 = 64$

$64 - 62 = 2$

There are 2 more girls than boys in the clubs.

Write/Explain: I **Wrote Number Sentences** from the data in the double-bar graph to find how many boys there are and how many girls there are in the clubs. I wrote each of the numbers of members of boys and girls in each club and added. There were 2 more girls than boys.

2. Graham Tries It. (page 118)

Score the Answer: I would give Graham a **2.** He gave the keywords and the facts, and he knew how to read the graph. He chose a good strategy and used it correctly. His interpretation of the graph was correct. He did not answer the question that was asked. It was not necessary to find the total number of students in the clubs.

Make it a 4! Rewrite.

Question: How many more are there of the group with more than the group with less?

64 − 62 = 2

There are 2 more girls than boys in the clubs.

Quiz Problems

Here are some problems for you to try. Keep your **rubric** handy while you solve the problem. Let's see if you can score a **4.**

1. Cal has scores of 83, 79, 92, and 87 on his math tests. He has another test tomorrow. What is the least score he can have to earn a mean score of 88?

2. There are 48 marbles in a bag. If there are 8 blue marbles, what is the probability that you will pick a blue marble at random?

3. There are eight teams playing in a single-elimination tournament. That means that if a team loses they are out of the tournament. One team will play another team in each game. How many games have to be played to determine a champion?

4. The temperature each day at noon for a week is listed in the table.

Day	Sun.	Mon.	Tue.	Wed.	Thu.	Fri.
Temperature in ° F	72	76	68	72	76	78

What is the median temperature at noon for the week?

5. Based on the graph below, how many people can be predicted to belong to the soccer club in 2006?

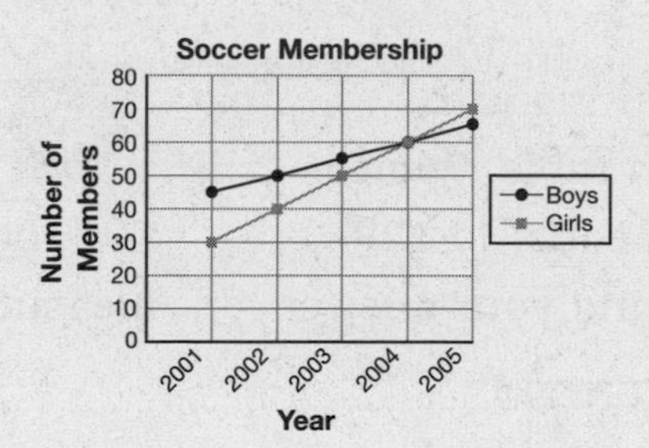

6. The bar graph shows the amount of money collected to help save a children's zoo from going out of business.

Stem	Leaves
0	3 8
1	0 3 5
3	6
4	2 2 5
7	0

Key: 1 | 3 = 13

How much money was collected all together?

7. Len bowled 4 games this weekend. His mean was 160, his median was 155, and his mode was 150. What was the range of Len's scores?

10. Test #1

Answer the following questions to the best of your ability. Remember, even if you are unsure of how to solve the problem, you will always earn some credit if you begin the problem. Good luck!

1. Maggie is sitting in the park, watching the people and dogs walk by. She counted 16 people and dogs combined and 38 legs. How many people did Maggie see walk by?

2. Nicole is working with this function table. What is the value of ***y*** if ***x*** is equal to 12? Explain how you found your answer.

x	*y*
2	2
4	8
7	17
10	26

3. Two of the angles of a parallelogram measure 70° and 110°. What are the angle measures of the other 2 angles?

4. A football field is 120 yards long. The field is also 160 feet wide. Before practice, the players run 5 laps around the field. How many feet more or less than a mile do the players run before practice?

5. Tino is going to toss a coin and a 6-sided number cube. What is the probability that he will toss tails and a number less than 3?

6. In a survey taken by the teachers at Callie's school, 4 out of every 5 students would rather go to an amusement park than a museum. If 75 students were surveyed, how many chose the amusement park?

10. Test #1

Answer the following questions to the best of your ability. Remember, even if you are unsure of how to solve the problem, you will always earn some credit if you begin the problem. Good luck!

7. Mrs. Hoffman gave this puzzle to her students. Solve her puzzle.
"Three times me plus 5 is equal to 26. What number am I?"

8. Hank plotted these ordered pairs on a coordinate grid: (1, 3), (3, 1), (5, 5), and (6, 3). What polygon did he draw? Answer as specifically as possible.

9. Candy and Mandy share a bedroom that is 18 feet long by 12 feet wide. They drew a diagonal from the southeast corner to the northwest corner of the room. Candy moved her stuff to one side of the diagonal. What is the area of Candy's side?

10. The cafeteria is offering a choice of hamburgers, pizza, or Caesar salad for an entrée, macaroni or mixed vegetables as a side dish, and bottled water, milk, or orange juice for a drink. How many choices of one entrée, one side dish, and one drink do students have?

11. Melissa walks $1\frac{1}{2}$ miles to school in the morning. After school, she walks $\frac{7}{8}$ mile to a park with her friends. How many more miles does Melissa walk to school than to the park?

12. For his first birthday, Justin received $1. For his second birthday, he received $3. For his third birthday, he received $6 and for his fourth birthday, he received $10. If the pattern continues, how much money will Justin receive for his tenth birthday?

10. Test #1

Answer the following questions to the best of your ability. Remember, even if you are unsure of how to solve the problem, you will always earn some credit if you begin the problem. Good luck!

13. Triangle *ABC* has ordered pairs of (4, 2), (4, 4), and (2, 1). Erin is going to reflect the triangle across the *y*-axis. What are the ordered pairs of the reflected image?

14. Mr. Page's car weighs 2,800 pounds. He drives to a bridge that says, "Weight Limit 2 Tons." Mr. Page, fearing for his life, turned the other way. Explain why Mr. Page was correct or incorrect in turning the other way.

15. It is Vanessa's goal to have an average of 95 in math. Her scores on the first five tests are 97, 92, 96, 89, and 100. What score does she need to get on her next test to have an average of 95?

16. A Girl Scout troop sold two times as many boxes of cookies Saturday as they did Friday. They sold 72 fewer boxes Sunday than they did Saturday. They sold 95 boxes of cookies Friday. How many boxes of cookies did they sell all together?

17. The school opened its doors at 6 p.m. for a 7 p.m. showing of the school play. There were 25 people who arrived at 6 p.m. Every ten minutes ***p*** people arrived until there were 145 people by 7 p.m. How many people arrived every ten minutes?

18. Tara drew as many diagonals as she could from one vertex of a regular hexagon. How many triangles did she form?

Answer the following questions to the best of your ability. Remember, even if you are unsure of how to solve the problem, you will always earn some credit if you begin the problem. Good luck!

19. A rectangular garden has an area of 72 square meters. The length of the garden is twice the width. What is the perimeter of the garden?

20. The circle graph shows the number of cast members in the school play by grade. If there are 40 people in the cast, how many are fifth-grade students?

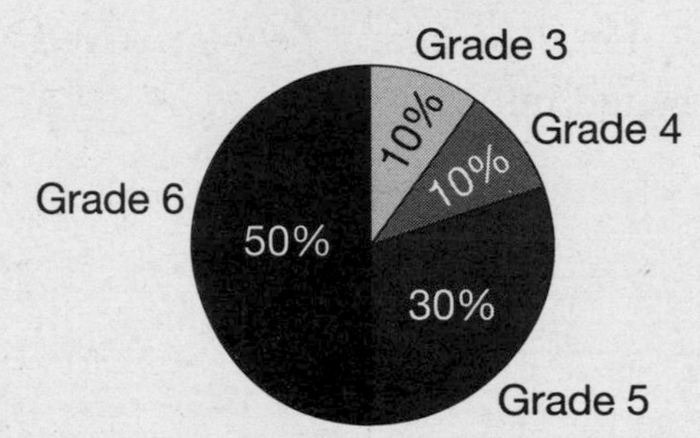

10. Test #2

Answer the following questions to the best of your ability. Remember, even if you are unsure of how to solve the problem, you will always earn some credit if you begin the problem. Good luck!

1. A football player has made 40 field goals this season. Field goals are worth either 2 points or 3 points. The player has scored 96 points from the field goals. How many 2-point field goals has the player made?

2. Sanjay is working with this function table. What is the value of y if x is equal to 12? Explain how you found your answer.

x	y
0	3
2	7
5	13
8	19

3. Two of the angles of an isosceles triangle measure 30° and 75°. What is the angle measure of the third angle?

Answer the following questions to the best of your ability. Remember, even if you are unsure of how to solve the problem, you will always earn some credit if you begin the problem. Good luck!

4. The gym at Teddy's school is 90 feet long and 48 feet wide. Before school, Teddy runs 8 laps around the gym. How many yards does Teddy run each morning?

5. Ashley is going to toss a coin and spin this spinner. What is the probability that she will toss tails and get a consonant?

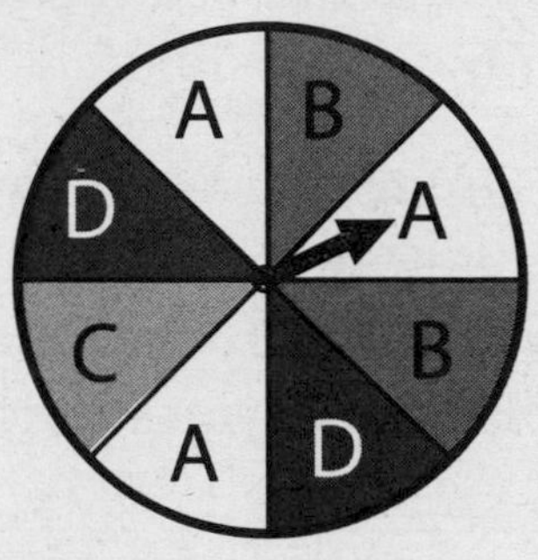

6. Sixteen out of the 24 students in Mrs. Jones's class wore sneakers to school. If the school has 360 people, how many students do you predict wore sneakers that day?

7. Mr. Morrow gave this puzzle to his students. "Four times me minus six is equal to 14. What number am I?"

8. Lara plotted these ordered pairs on a coordinate grid: (2, 1), (4, 1), (2, 5), and (4, 5). What figure did she draw? Answer as specifically as possible.

9. An island on Main Street is triangular. The base of the triangle is 20 yards long and the height of the triangle is 15 yards. What is the area of the island?

10. Test #2

Answer the following questions to the best of your ability. Remember, even if you are unsure of how to solve the problem, you will always earn some credit if you begin the problem. Good luck!

10. Randi wants to buy a matching top, skirt, and shoes for a party. She is choosing from a blue top, black top, red top, or striped top, a short skirt or a long skirt, and sneakers, sandals, or boots. How many choices of one top, one skirt, and one pair of shoes does Randi have?

11. Dmitri spent $2\frac{1}{4}$ hours writing a book report this weekend. He also spent $1\frac{1}{2}$ hours working on his math homework. How much more time did he spend writing his book report than working on his math homework?

12. What is the eighth number in this pattern?
2, 6, 14, 30, . . .

13. Rectangle ***WXYZ*** has ordered pairs of (1, 3), (1, 5), (4, 3) and (4, 5). Jeff is going to reflect the rectangle across the ***x***-axis. What are the ordered pairs of the reflected image?

14. For field day, there are races of 100 meters, 200 meters, 400 meters, 800 meters, and 1.5 kilometers. Marion plans to compete in all five races. How many kilometers will she have to run all together to compete in all five races?

15. It is Doug's goal to have a 150 bowling average. His average through 7 games is 147. What does he need to score in his eighth game to have an average of 150?

Answer the following questions to the best of your ability. Remember, even if you are unsure of how to solve the problem, you will always earn some credit if you begin the problem. Good luck!

16. A movie theater sold 3 times as many tickets on Friday as they did on Thursday. The theater sold 70 more tickets Saturday than they did on Friday. The theater sold 75 tickets on Thursday. If the theater sold 25 fewer tickets on Sunday than on Saturday, how many tickets were sold on Sunday?

17. Derek had 125 CDs when he joined a CD club. He purchases ***c*** CDs each month. After 7 months he has 167 CDs. How many CDs does Derek buy each month?

18. What is the angle measure of each angle of a regular pentagon? Explain your answer.

19. A right triangle has an area of 45 square inches. The base is 1 inch longer than the height. What are the dimensions of the right triangle?

20. The line graph shows the high temperature each day for 6 days. What is the median high temperature?

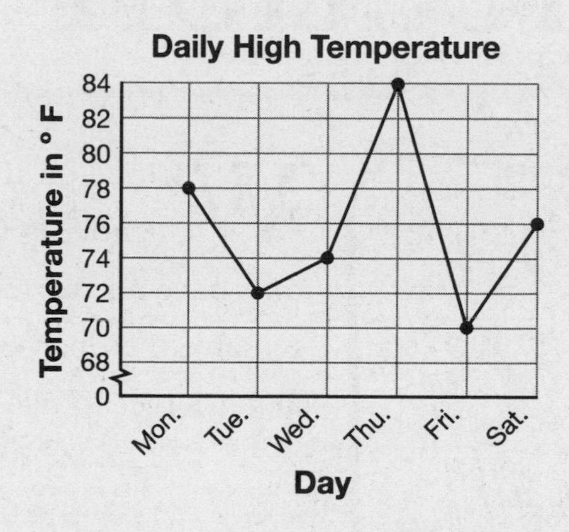

11. Home-School Connection

Working on these questions at home with a family member is fun! Find a comfortable place to work and have all the tools you need. Discuss how you want to solve the open-ended math question. Then go for it! Don't forget to use your rubric!

11. Home-School Connection

Dear Family Member:

This year your child will be learning about open-ended math questions in mathematics class. An open-ended math question is a mathematics word problem that has one correct answer, but that can be solved in several different ways. Open-ended math questions are extremely important on tests your child will take.

You can help your child practice solving these questions by working together on the take-home sheets in this chapter. Don't forget to use the rubric as a guide. Remember, when you work with your child, do not do the problem for him or her. Rather, encourage your child to ask questions that will lead him or her to the answer. And, do not be surprised if your child arrives at the answer to the problem using a method different from the one you're thinking of.

It is important that your child makes his or her thinking clear to the reader. After having solved the problem, (or when your child has gone as far as he or she can) help your child write a clear explanation of what he or she did to solve it, and why he or she decided to do it that way. This will help your child clarify his or her own thoughts.

The problems on the following pages are based on the areas of mathematics considered important in solving open-ended math problems. These are:

- **Number and Operations**
- **Algebra**
- **Geometry**
- **Measurement**
- **Data Analysis and Probability**

Enjoy!

Number and Operations

Problem

A truck driver has a route that takes him 125 miles each way each day that he works. If he makes 15 roundtrips on this route per month, how many miles does he drive the truck per month?

Algebra

Problem

Laura is working with this function table. The table shows what has happened so far.

x	*y*
5	8
8	14
13	24
16	30
19	?

What is the missing number? Explain how you found your answer.

Geometry

Problem

Besides a quadrilateral, what other ways can a square be described? Use as many as you can.

Measurement

Problem

Roz has a rectangular garden that is 10 meters by 8 meters. Next year she plans on doubling the dimensions of the garden. How will the perimeter and the area of next year's garden compare to this year's garden?

Data Analysis and Probability

Problem

The line graph shows the temperature each hour from 9 a.m. until 2 p.m. What is the median temperature for this period?

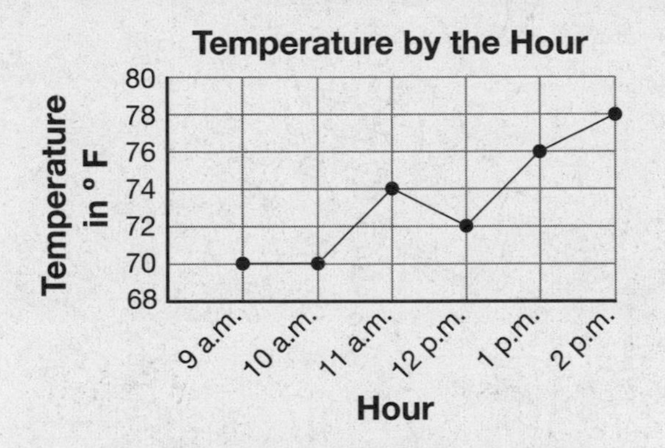

Glossary

A

After	At a later time.
Addend	One of two or more numbers that are added.
Algebra	Algebra uses letters and symbols to write expressions about number relationships.
All together	With all included.
Apart	A distance between two objects.
Area	The number of square units needed to cover a region.
At least	Equal to or greater than another number.
At random	A sample where everyone in a population has an equal chance of being picked.

Glossary

B

Bar graph A graph that shows data by using bars of different lengths.

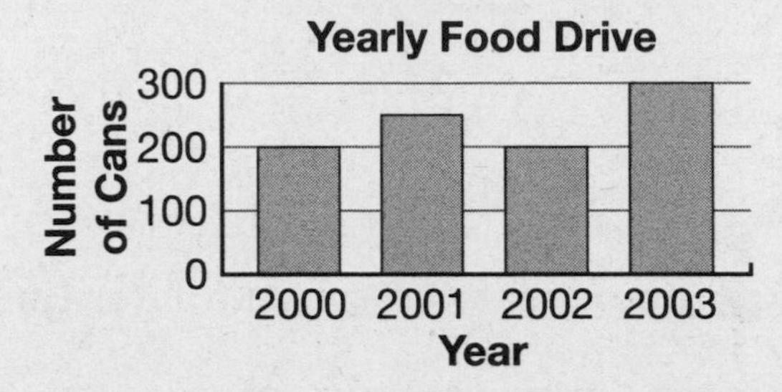

Base The side of a polygon, usually the one at the bottom.

C

Choice To select.

Circle graph A graph that uses a circle to show the parts of a whole.

Circular Having the shape of a circle.

Combined Joined.

Congruent A figure having the same size and shape as another figure.

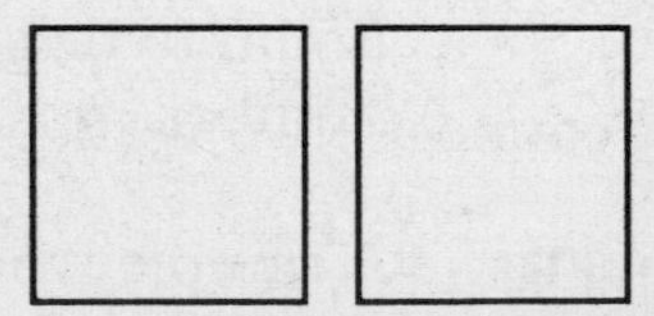

These squares are congruent.

Coordinate grid A grid used to show location.

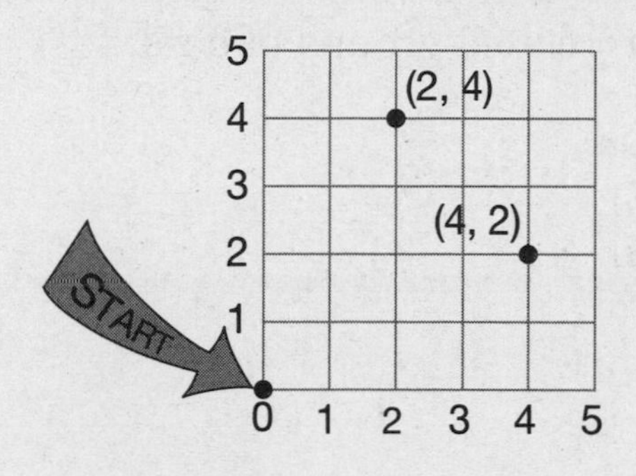

Cover To enclose.

Crosses Meets.

D

Deep Having depth.

Degrees Celsius (°C) System for measuring temperature where water freezes at 0° and boils at 100°.

Degrees Fahrenheit (°F) System for measuring temperature where water freezes at 32° and boils at 212°.

Depart To leave.

Diagonal A line segment other than a side that connects two vertices of a polygon.

Diameter A chord that passes through the center of a circle.

Digit Any of the numbers 0, 1, 2, 3, 4, 5, 6, 7, 8, or 9.

Dimensions The measures of a polygon or solid figure.

Double-bar graph A bar graph that compares two related groups of data.

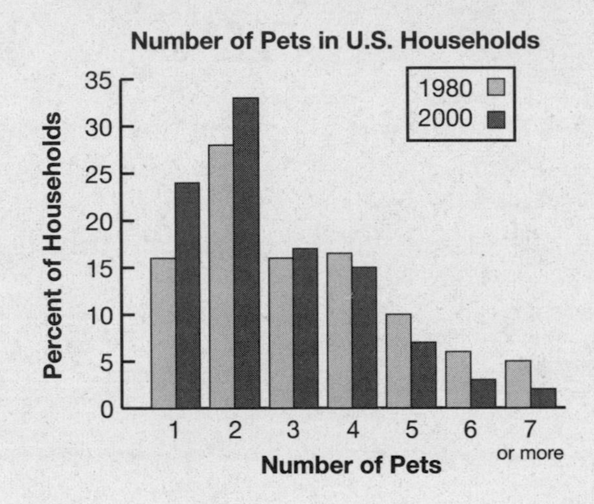

E

Each Every one of a group considered individually.

Either One or the other.

Every All.

F

Far A great distance.

Farther At greater distance than something else.

Fee A cost.

Fluid ounce A customary unit of capacity. 8 fluid ounces = 1 cup.

Foot A customary unit of length equal to 12 inches.

Form To create.

Formula An equation with at least two variables showing how one variable depends on the other variable or variables.

Function A relationship in which one quantity depends on another quantity.

G

Gallon A customary unit of capacity equal to 128 fluid ounces.

Geometry The study of figures.

Greatest The largest number in a set.

H

Half One of two equal parts that make a whole.

Height (High) Length.

Hexagon A polygon with 6 sides and 6 angles.

High A distance upward.

Hour One of 24 equal parts of a day.

Glossary

I

Image The figure made after a transformation.

Inch A customary unit of length equal to $\frac{1}{12}$ of a foot.

Increased To become larger.

Is equal to (=) A symbol used to show that two quantities have the same value.

Isosceles Triangle A triangle with 2 equal sides and two equal angles.

isosceles triangle

K

Kilometer A metric unit of length equal to 1,000 meters.

L

Lap Once around the perimeter or circumference.

Least The smallest number in a set.

Left Remaining.

Length The measurement of distance between two endpoints.

Line graph A graph that uses a line to show change over time.

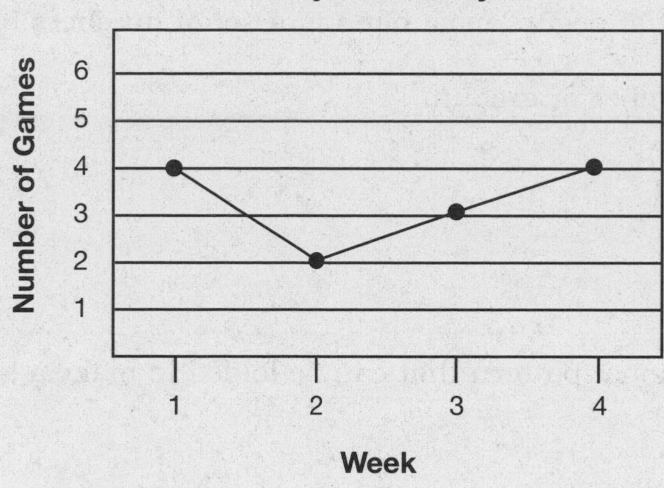

Line of symmetry A line on which a figure can be folded so that both sides match.

Long Length.

M

Many A large number or a certain amount.

Mean The quantity found by adding the numbers in a set of numbers and dividing their sum by the number of addends.

Median The middle number in an ordered set. If the set contains an even number of numbers, the median is the mean of the two middle numbers.

Meter A metric unit of length equal to 100 centimeters.

Mile A customary unit of length equal to 5,280 feet.

Minute A unit of time equal to 60 seconds.

Mode The number that occurs most often in a set of numbers.

More Greater in number or amount.

Most More than $\frac{1}{2}$.

Much An amount.

N

Net A two-dimensional pattern that can be folded to make a solid figure.

O

Old Of a certain age.

Older Something/someone existing for a longer time.

Open-ended problem A math problem that has a correct answer that you can arrive at in more than one way.

Ordered pair A pair of numbers that gives the location on a coordinate grid.

P

Parallel lines Lines that never meet and remain the same distance apart.

Parallelogram A quadrilateral in which both pairs of opposite sides are parallel.

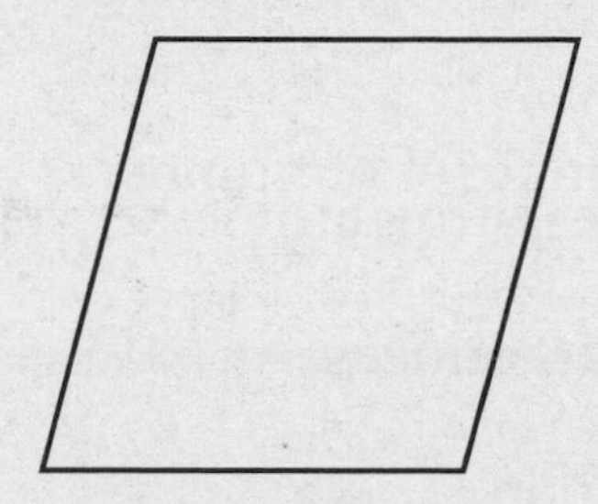

Pattern A series of numbers or figures that follows a rule.

Pentagon A shape with 5 sides and 5 interior angles.

Per For each.

Percent Per hundred or part of a hundred.

Perimeter The distance around a figure. The perimeter is found by adding the lengths of all of a polygon's sides.

Perpendicular lines Lines that meet at a right angle.

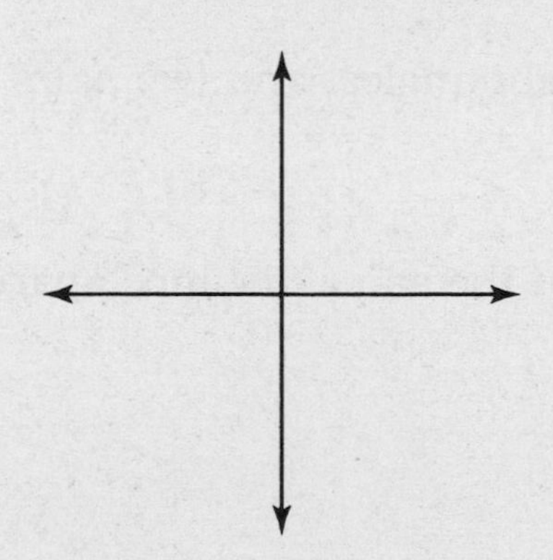

Glossary

Pi (π)	The ratio of the circumference of a circle to its diameter. It is estimated as 3.14.
Plotted	Drew.
Polygon	A closed geometric figure with all sides straight line segments.
Pound	A customary unit of weight equal to 16 ounces.
Predict	To foretell on the basis of an observation.
Prime number	A whole number greater than 1 that has only 1 and itself as factors.
Probability	The chance of an event occurring.

Q

Quadrilateral	A polygon with four sides and four angles.
Quart	A customary unit of capacity equal to 32 fluid ounces.

R

Range	The difference between the greatest and the least numbers in a data set.
Rectangle	A polygon with four sides, four right angles, and two pairs of opposite congruent sides.

Rectangular Having the shape of a rectangle.

Reflect A transformation that creates a mirror image of a figure across a line.

Regular A polygon with all sides the same length.

Remaining Left.

Returned Having come back to the original location.

Rhombus A parallelogram with four congruent sides.

Right angle An angle that forms a square corner.

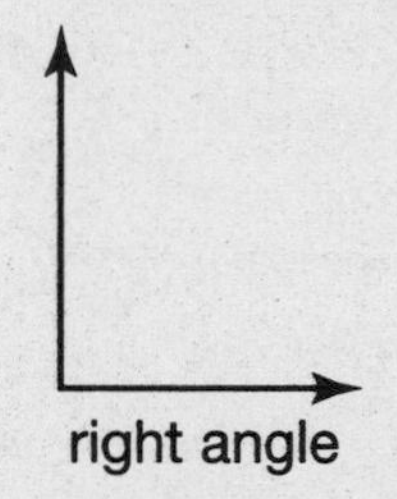

Glossary

Right triangle A triangle with one right angle.

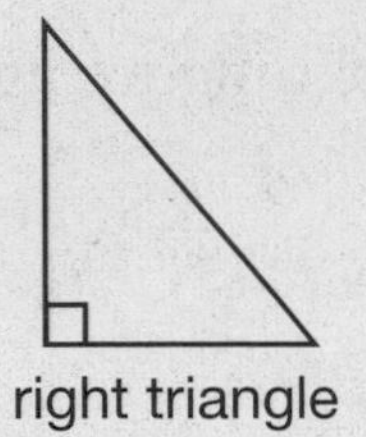

right triangle

S

Same Being equal in quantity or quality to another item.

Similar Figures that have the same shape, but may have different sizes.

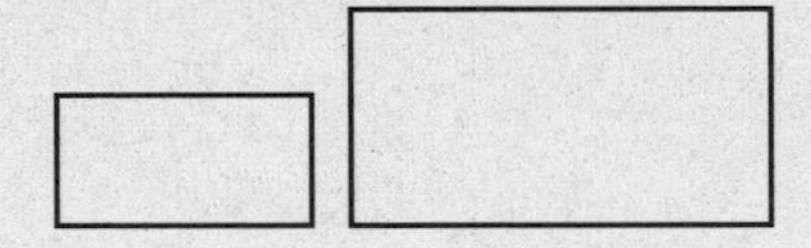

These squares are similar.

Specifically As exact as possible.

Square A rectangle with four equal sides.

Square yard	A square that measures 1 yard on each side.
Successful	Achieving the desired result.
Surface area	The total area of the surface of a solid figure.
Survey	A method used to gather data that involves asking people questions or observing events.

T

Temperature	A measurement that tells how hot or cold something is.
Tessellation	An arrangement of shapes that covers a region without any gaps or overlaps.
Three-dimensional figure	A figure that has length, width, and height.
Times	Multiplied by.
Tip	A sum of money given for a service.
Ton	A customary unit of weight equal to 2,000 pounds.
Triangle	A polygon with three sides and three angles.
Twice	Two times.

V

Value	An assigned or calculated quantity.

W

Wide	Width.
Width	The length of the shorter sides of a polygon.

X

***x*-axis**	The horizontal axis of a coordinate grid.

Y

***y*-axis**	The vertical axis of a coordinate grid.
Yard	A customary unit of measure equal to 3 feet.
Younger	Someone/something existing for a shorter time.

Math Abbreviations

centimeter	cm	hour	h	milligram	mg	pound	lb
cup	c	inch	in.	milliliter	mL	quart	qt
day	d	kilogram	kg	millimeter	mm	second	s
fluid ounce	fl oz	kilometer	km	minute	min	ton	T
foot	ft	liter	L	month	mo	week	wk
gallon	gal	meter	m	ounce	oz	yard	yd
gram	g	mile	mi	pint	pt	year	y

Math Symbols

$+$	addition	"	inches
$-$	subtraction	'	feet
$\times$	multiplication	°C	degrees Celstus
$\div$	division	°F	degrees Fahrenheit
$=$	is equal to	$\overleftrightarrow{AB}$	line AB
$>$	is greater than	$\overrightarrow{AB}$	ray AB
$<$	is less than	$\overline{AB}$	line segment AB
.	decimal point	$\angle A$	angle A
$	dollars	$\triangle ABC$	triangle ABC
¢	cents	[2, 3]	ordered pair [2, 3]